AN OVERVIEW OF GRAVITATIONAL WAVES

History Sources Properties and Detection

SANYASI RAJU SAGI

notionpress
.com

INDIA • SINGAPORE • MALAYSIA

ISBN

Paperback 979-8-89673-416-1
Hardcase   979-8-89929-852-3

Dedication

This book is dedicated to the precious memory of the philanthropist and visionary Dr PVG Raju, Raja Saheb of Vizianagaram, who made immense contributions in the field of education through his brainchild MANSAS, under the canopy of which I am fortunate enough to render my services for 38 years.

Synopsis

Albert Einstein's 'General Theory of Relativity' is a masterpiece of Mathematics and Physics entwined in a spectacular mosaic of spacetime that engulfs the whole Universe. One direct outcome of this incredible theory is the generation of gravitational waves due to the collision and merger of astronomical behemoths: Black holes and Neutron stars. This book **"An Overview of Gravitational Waves: History, Sources, Properties and Detection"** is an attempt to highlight the fascinating discovery of gravitational waves and the saga of several scientists and researchers who worked for nearly six decades to nail this discovery. The major role played by the engineering marvel Laser Interferometer Gravitational Wave Observatory (LIGO) with its associated companions VIRGO and KAGRA in this historic detection of gravitational waves is explained in the book with their principles, design, and working conditions in an elaborate manner. Not lagging behind in this eventful adventure, our Indian scientists' contribution and collaboration efforts along with the establishment of the third LIGO observatory on our Indian soil is another chapter incorporated in this book.

While this book is primarily for those students and researchers who are enthusiastic about keeping themselves up to date with these trend-setting developments in the realm of astrophysics and General Relativity, the author feels this book gives a sense of exhilaration to all the ardent lovers of physics.

Contents

Preface

This book, the first in the series of monographs entitled **"The Frontiers of Physics"**, is intended to convey the excitement of pioneering discoveries in recent times in the realm of the microscopic as well as macroscopic worlds of cosmology, relativity, and high-energy physics.

On 14 September 2015, the LIGO detectors in the USA saw space vibrate with gravitational waves for the very first time. Although the signal was extremely weak when it reached Earth, it is already promising a revolution in astrophysics. Gravitational waves are an entirely new way of following the most violent events in space and testing the limits of our knowledge. This signalled the beginning of the biggest change in astronomy since the introduction of telescopes, and our understanding of the Universe took a leap forward.

Einstein predicted that something special happens when two bodies—such as planets or stars—orbit each other. He believed that this kind of movement could cause ripples in space. These ripples would spread out like the ripples in a pond when a stone is tossed in. Scientists call these ripples of space—**gravitational waves**.

Gravitational waves are created by some of the most powerful and violent events in the Universe. Detecting gravitational waves on Earth was a challenging task that took roughly a century to complete since the ones that wash through the planet are incredibly tiny.

The first indirect evidence for the existence of gravitational waves came in 1974 from the observed orbital decay of the Hulse–Taylor binary pulsar, which matched the decay predicted by General Relativity as energy is lost to gravitational radiation. Russell A Hulse and Joseph Hooton Taylor Jr received the Nobel Prize in Physics in the year 1993 for this discovery. The 2017 Nobel Prize in Physics was subsequently awarded to Rainer Weiss, Kip Thorne, and Barry Barish for their role in the direct detection of gravitational waves.

Gravitational waves are invisible. However, they are incredibly fast. They travel at the speed of light (186,000 miles per second). Gravitational waves squeeze and stretch anything in their path as they pass by. When two black holes inspiral and merge, a significant portion of their mass can get converted into energy in one very short time interval. The gravitational radiation that these black holes emit doesn't propagate across the Universe instantaneously but rather is limited by the finite speed of propagation: the speed of light. Although gravitational waves were deduced from Einstein's General Relativity all the way back in 1915, it took a hundred years for humanity to successfully detect them. Even though we've detected merging black holes, merging neutron stars, and neutron stars merging with black holes via gravitational waves, so much more is still to come.

Numerous detection schemes have been proposed, including vibrating bars that would be sensitive to the oscillatory motion of a passing gravitational wave, pulsar timing that would be sensitive to oscillatory changes of gravitational waves that passed through the pulse's line of sight with respect to us, and reflected laser arms that span different directions, where the relative changes between the multiple path lengths would reveal the evidence of a gravitational wave as it passed through. The last of these is precisely the first—and thus far, the only— method by which we've ever successfully detected gravitational waves. Since that time, the twin LIGO detectors have been joined by two other ground-based Laser Interferometer gravitational wave detectors: the VIRGO detector in Europe, and the KAGRA detector in Japan. Later this decade, they'll be joined by a 5th detector, LIGO-India, which will increase their sensitivity even further.

The research and discovery of gravitational waves by the LIGO observatory are truly exciting and intriguing, and it surely fires the imagination of researchers and scientists all over the world. If the traditional astronomy of telescopes is like 'seeing' the cosmos, then gravitational wave astronomy is akin to 'hearing' it. Gravitational Waves even provide a window into what happened in the first few moments after the Big Bang. The Universe is home to some amazing things! Gravitational waves are an entirely new way of following the most violent events in space and testing the limits of our knowledge.

Writing a book of this nature, wide in scope and reasonably concise, requires a lot of effort that includes going through resource material from institutions and organisations like NASA, National Science Foundation (NSF), LIGO, LISC, Inter-University Centre for Astronomy and Astrophysics (IUCAA), and publications in journals such as Scientific American, Nature, Symmetry, Physics Today, and Resonance. I acknowledge my indebtedness to all those institutes and researchers whose work I extensively used in making this book see the light of day.

Writing this book also reflects my belief that science writing on a topic of gigantic discovery of this nature like gravitational waves, with the goal of attaining a wider readership, has a place in modern book publishing. I hope this publication will rekindle the spirit of human ingenuity in the minds of young students and scholars of higher learning.

"An Overview of Gravitational Waves: History, Sources, Properties and Detection" has been an arduous project, brought to fruition through the efforts of some very special people. I am greatly indebted to my mentors at the Department of Nuclear Physics, Andhra University, Visakhapatnam, and convey my regards to Prof B Mallikarjuna Rao and Dr B Seetha Rama Reddy for nurturing my doctoral research. I am very grateful to all my teachers for their guidance while pursuing my postgraduate and doctoral research at Andhra University, including Prof V Laxminarayana, K Parthasarathi, MLN Raju, Prem Chand, Bhooloka

Reddy, Seshi Reddy, and last but not least Prof MVR Murty of IIT, Mumbai. I convey my special thanks to my research colleagues: Ram Narayana, Naga Raju, Abdul Sattar, and Koteswara Rao.

The genesis for writing this book goes to my solitude time spent in the Dr VS Krishna Memorial Library at Andhra University in the late 1970s where I could lay my hands on popular scientific journals of the kind: Scientific American, Physical Review Letters, and Physics Today. The elaborate discussions I had with my friend and philosopher Dr A Raghava Rao and my friendly chatter with my good friends and colleagues at MR College, Vizianagaram—Dr DRK Raju, Dr MA Kareem, and Mr P Chandra Rao—are invigorating and memorable. I convey my sincere thanks to all my former colleagues of the Department of Physics, MR College: Sri GV Surya Narayana, DB Subba Rao, D Jaganadha Rao, A Venkata Rao, GV Ramana Murthy, PV Appa Rao, D Sanyasi Rao, Sitarama madam, DBRK Murthy and also to physics faculty members of MRPG College, Vizianagaram for their great support. My special thanks to Jayapal Rao who has given valuable inputs for the publication of this book.

I convey my special thanks to all my former colleagues at Maharajah's College, Vizianagaram for their support and guidance while writing this book, including Achuta Rao, Krishna Kishore, MDP Patnaik, Narasimha Rao, Siva Rama Murthy, Rama Raju, PAN Raju, Padmanabha Raju, PSN Rao, and SGK Sharma. My regards to all my friends at IAPT: RC-11 and especially to J Chandrasekhar.

My final words of thanks are reserved for former Chairmen of MANSAS Trust, Sri P Ananda Gajapathi Raju and Sri P Ashok Gajapathi Raju, and correspondents, Sri PLN Raju, R DSN Raju, and KVL Raju for their blessings and support in my journey of three and a half decades of work in MANSAS. And of course, to my family members: my parents Bangar Raju and Sita Devi, my wife Aruna, my son Santosh, his wife Samyuktha, and my granddaughter Sri Vindhya for their tolerance and forgiveness and for their help in keeping the work-life balance stable despite lapses from my side.

The purpose of this book is to present the history and recent developments in the detection of gravitational waves and their characteristic and exotic properties in an unpretentious manner for a better understanding of the subject without going into the nitty-gritty of equations. The complex mathematics of Hilbert Spaces, Wave functions, Tensors, and Operators underpinning General Relativity can intimidate readers and drive them away from the exciting world of gravitational waves. The key readership for this book is an interested general audience as well as students interested in an overview of gravitational waves.

The book is expected to make it a pleasure cruise as well as a voyage of discovery.

CHAPTER – 1

Introduction

"In Science, often the grandest discoveries are triggered by the smallest discrepancies."

The discovery of gravitational waves is no exception! Gravitational waves are distortions in the fabric of space and time caused by the movement of massive objects like stars, black holes, neutron stars, and supernova explosions. Like ripples from a pebble thrown into the water, gravitational waves from these cosmic explosions spread through the fabric of the cosmos.

These waves are similar to sound waves in the air or the ripples made on a pond's surface when someone throws a rock in the water. But unlike waves of pond ripples, which spread out through a medium like water, gravitational waves are vibrations in spacetime itself, which means they move just fine through the vacuum of space. And unlike the gentle drop of a stone in a pond, the events that trigger gravitational waves are among the most powerful in the Universe.

We can hear gravitational waves, in the same sense that sound waves travel through water, or seismic waves move through the Earth. The difference is that sound waves vibrate through a medium, like water or soil. For gravitational waves, the spacetime fabric itself is the medium. It just takes the right instrument to hear them.

The research and discovery of gravitational waves is truly exciting and intriguing, and it surely fires the imagination of thousands of scientists throughout the world.

GENERAL RELATIVITY

Einstein's General Theory of Relativity, published in November 1915, led to the prediction of the existence of gravitational waves. He himself wondered whether they could ever be discovered as they were not only faint but also their interaction with matter was abnormally weak. Even if they were detectable, Einstein also doubted if they would ever be useful enough in scientific applications.

General Theory of Relativity replaced Newtonian gravity by treating spacetime as a four-dimensional fabric where all the matter and energy travelled through that fabric limited by the speed of light. That fabric wasn't simply flat, like a Cartesian grid, but rather had its curvature determined by the presence and motion of matter and energy: **matter and energy tell spacetime how to curve, and that curved spacetime tells matter and energy how to move**. Einstein's general theory of relativity offered a profound recasting of gravity in terms of a startling new idea: warps and curves in space and time. In our everyday lives, we think of 3-dimensional space (up/down, left/ right, forward/backwards) and time as completely separate things. But Einstein's Theory of Relativity showed that the three spatial dimensions plus time are actually just part of the same thing: the four dimensions of spacetime.

The distortions of space and time are most noticeable in the vicinity of large masses or compact objects. Imagine that spacetime is like a giant rubber sheet. When a massive object (like a star) sits on the sheet, the sheet (and therefore spacetime) bends and becomes curved, and anything placed near the large object will accelerate and fall towards the massive object. From Einstein, we know that gravity works a lot like this. When we experience a gravitational force accelerating us towards the Earth, we're actually travelling along curves in space and time!

The Theory of General Relativity predicts a number of phenomena, including the bending of light by gravity (gravitational lensing), the gravitational redshift, black holes, the existence of which has been confirmed by observation and experiment, and of course gravitational waves.

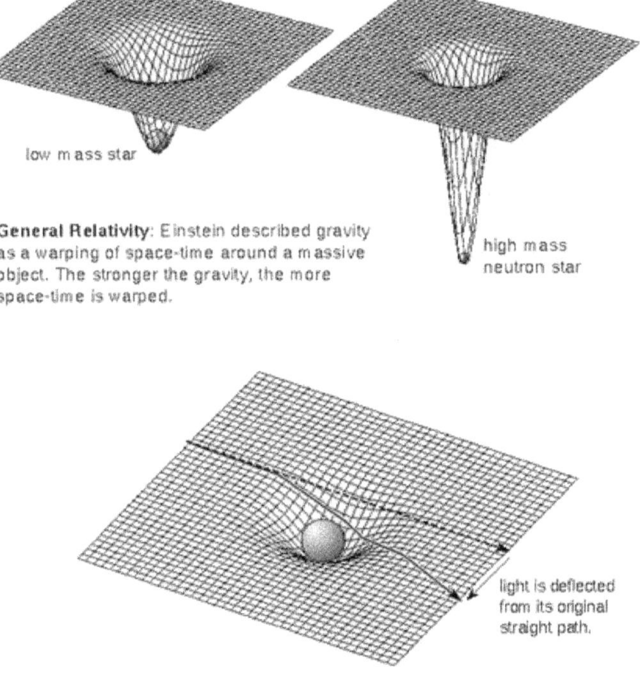

Figure 1-1.   Spacetime distortion due to heavy objects such as stars.

Gravitational lensing occurs when a massive celestial body such as a galaxy cluster causes a sufficient curvature of spacetime for the path of light around it to be visibly bent, as if by a lens. The body causing the light to curve is accordingly called a gravitational lens. Gravitational lensing is a dramatic and observable example of Einstein's theory in action. As shown in the figure below, the light can be seen to bend around the body.

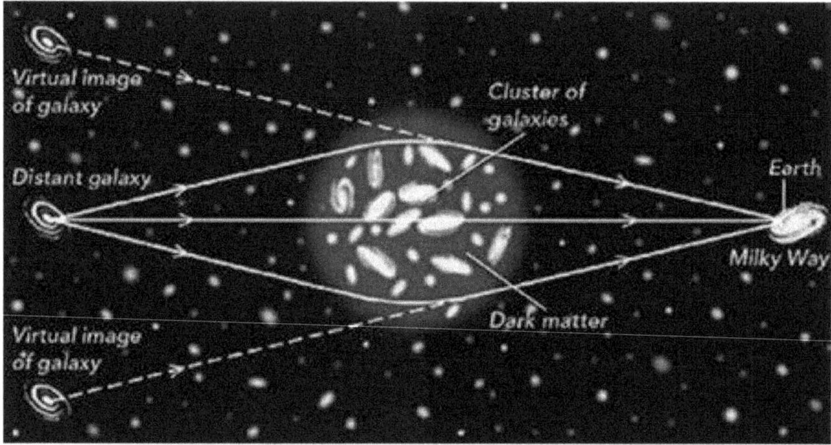

Figure 1-2.   Gravitational lensing effect of a galaxy.

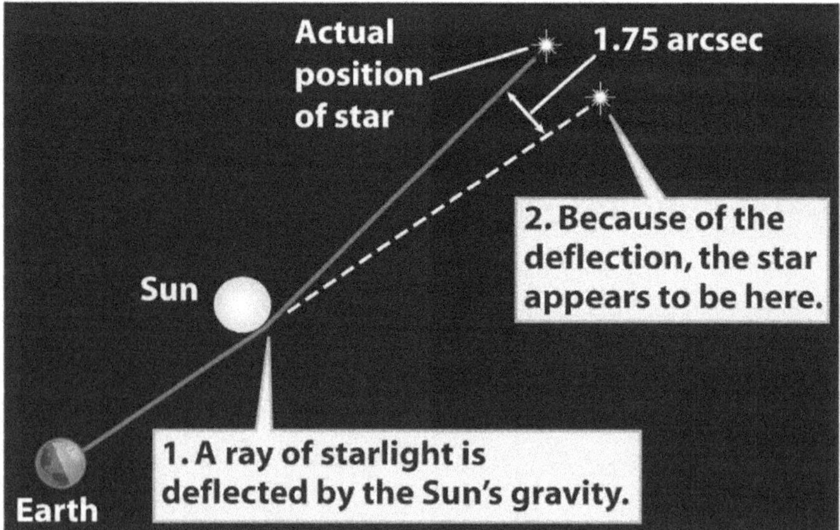

Figure 1-3.   Gravitational lensing effect due to the Sun.

BLACK HOLE

Black Holes are the most profound prediction of General Relativity. A black hole is a large body of matter that is so dense that nothing can escape its gravitational attraction, at a given distance, known as the Schwarzschild radius. If a stellar corpse has a mass greater than about 2 to 3 M_0 (Solar mass), gravitational compression will overwhelm any and all forms of internal pressure. The stellar corpse will collapse to such a high density that its escape speed exceeds the speed of light and thus forms a black hole. Black holes are famous for devouring any object that crosses their event horizons. At the same time, they are thought to continuously leak energy to space in the form of Hawking radiation. Proposed by Stephen Hawking in 1974, this emission is black-body radiation that causes a black hole to shrink and eventually disappear.

Astrophysicists distinguish mainly between the following three types of black holes: (i) Stellar black holes, which have masses of about 3 to 30 solar masses and are formed due to the exhaustion of thermonuclear fuel in them and the subsequent supernova explosion; (ii) Intermediate mass black holes, which have masses of 100 –1000 solar masses and form from the merger of small black holes in the centres of dense stellar clusters or from the death of very massive stars early in the history of the Universe; and (iii) massive black holes of mass 10 thousand – 10 billion solar masses, which are found in the centres of most galaxies. Our Milky Way also has such a black hole at its centre, whose mass has been estimated to be 4 million solar masses.

A dip in spacetime arises in the gravitational field (GF). The GF of one object can affect another object. The object of smaller mass might fall into the object of heavier mass's GF and orbit around it, like the Moon around the Earth, or Earth around the Sun. Alternatively, two bodies with GFs might spiral towards each other, getting closer and closer until they collide. As this happens, they create ripples in spacetime— gravitational waves.

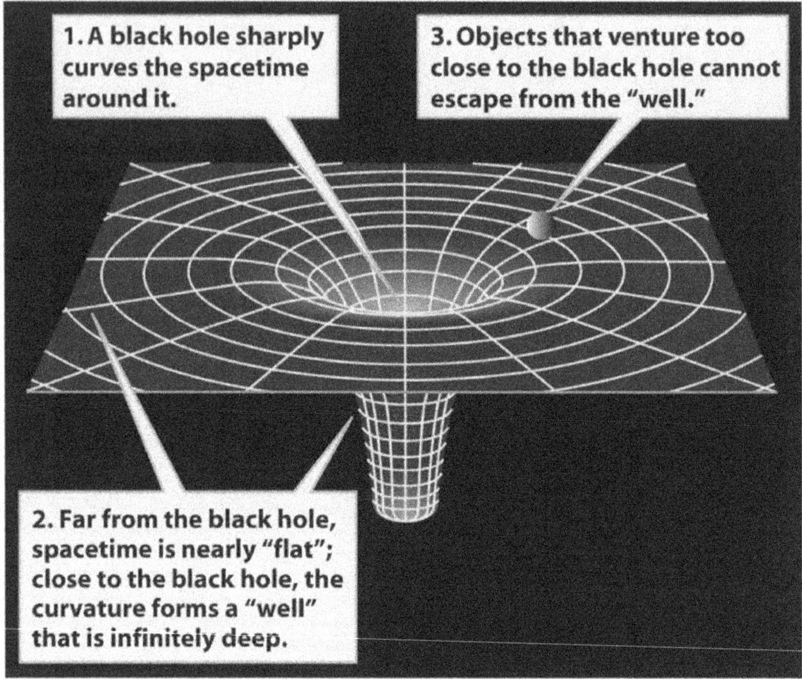

Figure 1-4.   Schematic of a Black Hole

Whenever an energy-containing object moves through curved space, one inevitable consequence is that it emits energy in the form of gravitational radiation, i.e. gravitational waves. They're everywhere in the Universe, and now that we've begun to detect them, they're about to open up new frontiers of astronomy.

Gravitational waves are very subtle and hold the potential of the unknown. Every time humans have opened new "eyes" to the Universe, we have discovered something unexpected that revolutionised how we see the Universe and our place within it. We are ushering into a new era with a new set of eyes that do not depend on light. Scientists are comparing the ability to directly detect gravitational waves to a deaf person suddenly gaining the ability to hear sound.

Exactly 100 years after Einstein's General Relativity (GR) was born, these waves were finally detected and are going to provide scientific results. In fact, at 11:50:45 am CET (Central European Time) corresponding to 15:20:45 pm IST, on 14 September 2015, Marco Drago—a postdoc—was seated in front of a computer monitor at the Max Planck Institute for Gravitational Physics in Hanover, Germany, when he received an email, automatically generated three minutes before from the monitors of LIGO (Laser Interferometer Gravitational Wave Observatory). Marco opened the email, which contained two links. He opened both links and each contained a graph of a signal similar to that recorded by ornithologists to register the songs of birds. One graph came from a LIGO station located at Hanford, in Washington State, and the other from Livingston Station in the state of Louisiana in the USA.

Marco is a member of a team of 30 physicists working in Hanover, analysing data from Hanford and Livingston. Marco's duty is to be aware of and analyse the occurrence of an "event" that records the passage of a gravitational wave, in one of the four lines that automatically track the signals from the detectors in the two LIGO observatories on the other side of the Atlantic. Marco noticed that the two graphs were almost identical, despite having been registered independently in sites separated by nearly 3000 km.

The time that elapsed between the two signals differed by about 7 milliseconds. These almost simultaneous records of signals coming from sites far away from each other, the similarity of their shapes and their large size, could not be anything but either a possible record of a gravitational wave travelling at the speed of light or a "signal" artificially "injected" to the detectors by one of the four members of the LIGO programme who are allowed to "inject" dummy signals. The reason why artificial signals are injected into the system is to test whether the operation of the detectors is correct and if the duty observers are able to identify a real signal. Following the pre-established protocols, Marco tried to verify whether the signals were real or the "event" was just a dummy-injected signal.

Since LIGO was still in engineering mode, there was no way to inject fake signals, i.e., hardware injections. Therefore, everyone was nearly one hundred per cent certain that this was a detection. However, it was necessary to go through the protocols to make sure that this was the case. Marco asked Andrew Lundgren, another postdoc at Hanover, to find out if the latter was the case. Andrew found no evidence of a "dummy injection". On the other hand, the two signals detected were so clear they did not need to be filtered to remove background noise. They were obvious. Marco and Andrew immediately phoned the control rooms at Livingston and Hanford. It was early morning in the United States, and only someone from Livingston responded. There was nothing unusual to report. Finally, one hour after receiving the signal, Marco sent an email to all collaborators of LIGO asking if anyone was aware of something that might cause a spurious signal. No one answered the email. Days later, LIGO leaders sent a report stating that there had not been any "artificial injection". By then, the news had already been leaked to some other members of the world community of astrophysicists. Finally, after several months, the official news of the detection of gravitational waves was given at a press conference on 11 February 2016, after the team had ascertained that the signals were not the result of some experimental failure, or any signal locally produced, earthquake, or electromagnetic fluctuation.

Now known as GW150914 (for the date on which the signals were received in 2015, 9th month, day 14), the event represents the coalescence of two black holes that were previously in mutual orbit. In the binary black hole merger, they needed to have started their mutual orbit separated by less than one-quarter of the distance between the Earth and the Sun. If the two black holes started out any farther apart, it would take longer than the age of the Universe for them to merge.

The LIGO observatories are funded by the National Science Foundation (NSF) and were conceived, built, and operated by Caltech and the Massachusetts Institute of Technology (MIT). The discovery, accepted for publication in the journal Physical Review Letters, was made by the LIGO Scientific Collaboration (which includes the GEO Collaboration and the Australian Consortium for Interferometric Gravitational Astronomy) and the VIRGO Collaboration using data from the two LIGO detectors.

LIGO's exciting discovery provides direct evidence of what is arguably the last major unconfirmed prediction of Einstein's General Theory of Relativity. This announcement is the most important scientific news so far in this century. This discovery not only confirmed one of the most basic predictions of General Relativity but also signalled the beginning of the biggest change in astronomy since the introduction of telescopes. Our understanding of the Universe took a leap forward. This discovery is significant since it makes it possible to see further deep into the Universe at objects that have been invisible so far.

Gravitational waves are ripples in spacetime produced by some of the most violent events in the cosmos, such as the collisions and mergers of massive compact stars or black holes. The two black holes orbited around a shared centre of mass at a significant percentage of the speed of light and, as they ploughed through spacetime, they threw out gravitational waves. But gravitational waves can't be made for free—it takes energy to churn up the fabric of the Universe—and, with each orbit, the black holes lost orbital energy, which was carried away by the gravitational waves, and their orbit shrank. As their orbit shrank, the black holes accelerated, which created more powerful gravitational waves, causing their orbit to shrink more, and the black holes to accelerate even more. It was a vicious circle that could only end one way: the collision of black holes. When they collided, they merged together and formed a single, even more, massive black hole, but the mass of the new black hole was less than the sum of its parts, as the mass difference is utilised in the formation of gravitational waves. All of this missing mass had been converted into gravitational energy. These ripples travel at the speed of light through the Universe, carrying with them information about their origins. LIGO's detectors are tuned to measure the gravitational waves produced by the system during the last phase of coalescence.

A "normal" system, like the Earth orbiting the Sun, will survive for an extraordinarily long time before enough gravitational waves are emitted to cause an inspiral and merger: something like 100 billion years for our planet, or much, much longer than the lifetime of the Sun itself. But for more massive systems at much closer distances, like binary neutron star systems or binary black hole systems, inspirals and mergers can take place in much shorter periods of time: timescales that are less than 10 billion years, or younger than the age of the Universe.

More than 1000 scientists work on the \$1-billion LIGO experiment, which is funded by the National Science Foundation (NSF). Each of the LIGO instruments is an engineering marvel, a laser-based measuring device capable of detecting a twitch in the width of an atom. LIGO works by shooting laser beams down two perpendicular arms, each 4 km long, and measuring the difference in length between them as the passing gravitational waves distort the spacetime fabric—a strategy known as laser interferometry. The gravitational waves that LIGO detected distorted the geometry of spacetime by just 10^{-18} metres—a thousandth the width of the nucleus of an atom—and it registered here on Earth as a brief and gentle chirp. LIGO is one of the most complex systems ever built by mankind. The experiment is so delicate that unrelated events such as an aeroplane flying overhead, wind buffeting the building, or tiny seismic shifts in the ground beneath the detector can disturb the lasers in ways that mimic gravitational signals. The researchers carefully weed out such contaminating signals and also take advantage of the fact that the detectors in Washington and Louisiana are highly unlikely to be affected by the same contamination at the same time.

In essence, LIGO is a celestial earpiece, a giant microphone that listens for the faint symphony of the hidden cosmos. Starting with LIGO's historic first detection in 2015, most of the 90 or so gravitational-wave events recorded so far have been from the spiralling motion of pairs of black holes in the process of merging into one; a handful has been produced similarly by the merger of two neutron stars or a neutron star and a black hole. Apart from LIGO, Advanced VIRGO in Italy, KAGRA in Japan, and a possible 3rd LIGO detector in India will extend the network and significantly improve the position, reconstruction, and parameter estimation of sources.

LIGO, VIRGO, and KAGRA are all based on the same interferometer concept, which involves splitting a laser beam into two and bouncing the resulting beams between two mirrors at either end of a long vacuum pipe. At LIGO, the two 'arms' of the interferometer are each 4 km long; at VIRGO and KAGRA, they are 3 km. The two beams then come back and are made to overlap at a sensor in the middle. In the absence of any disturbances to spacetime, the beams' oscillations cancel each other out. However, the passage of gravitational waves causes the arms to change in length with respect to each other so that the waves don't overlap perfectly, and the sensor detects a signal.

The discovery is not just proof of gravitational waves, but the strongest confirmation yet for the existence of black holes. Given that black holes themselves cannot give any signal other than gravitational waves, this is the most direct way to prove that a black hole exists. Detecting and analysing the information carried by gravitational waves will allow us to observe the Universe in a way never before possible. This will open up a new window of study on the Universe, giving us a deeper understanding of these cataclysmic events, and usher in brand new cutting-edge studies in physics, astronomy, and astrophysics.

Historically, scientists have relied primarily on observations with electromagnetic radiation (visible light, X-rays, radio waves, microwaves, etc) to learn about and understand phenomena in the Universe. In addition, some types of subatomic particles, including neutrinos and cosmic rays, have been used to study cosmic objects. Scientists build different types of telescopes and observatories to detect these various signals. Some telescopes are located on the ground (e.g., visible light and radio), some are located on satellites orbiting Earth (such as X-ray and gamma-ray, as well as the Hubble Space Telescope which images visible light and ultraviolet), and still others are buried in ice (neutrino detectors) or are far underground (particle detectors). Each provides a different and complementary view of the Universe, with each new window bringing exciting new discoveries.

CHAPTER – 2

History

On 5 July 1905, the French Academy of Sciences published an article written by Henri Poincaré on his Theory of Relativity, which summarised that gravity was transmitted through a wave called a gravitational wave. Poincaré may have been the first to use the term gravitational wave. He speculated about (special) relativistic gravity, which would have waves of acceleration, a retarded attractive force that propagated at the speed of light. The first seed in the discovery of gravitational waves was sown then.

It would take some years for Albert Einstein to postulate in 1915 in final form the Theory of General Relativity. His theory can be seen as an extension of the Special Theory of Relativity postulated by him 10 years earlier in 1905. The General Theory explains the phenomenon of gravity. It is nothing less than the key to understanding the history of the Universe, the origin of time, and the evolution of all the stars and galaxies in the cosmos. In this theory, gravity is not a force—a difference from Newton's Law—but a manifestation of the curvature of spacetime, this curvature being caused by the presence of mass and also energy and momentum of an object. In other words, Einstein's equations are able to match the curvature of local spacetime as well as the local energy and momentum within that spacetime. Space is no longer just a space where things exist, nor is time a ticking clock keeping tabs on things. According to Einstein, space and time are intertwined in a cosmic dance encompassing all the stuff imaginable, from particles to galaxies, weaving themselves into elaborate patterns that can lead to the most bizarre effects in the Universe.

Einstein's Enigma

Shortly after finishing his theory, Einstein conjectured, just as Poincaré had done, that there could be gravitational waves similar to electromagnetic waves. The latter are produced by accelerations of electric charges. In the electromagnetic case, what is commonly found is dipolar radiation produced by the swinging of an electric dipole. An electric dipole is formed by two (positive and negative) charges that are separated by some distance. Oscillations of the dipole separation generate electromagnetic waves. However, in the gravitational case, the analogy breaks down because there is no equivalent to a negative electric charge. There are no negative masses. In principle, the expectation of theoretically emulating gravitational waves similar to electromagnetic ones faded in Einstein's view.

However, Einstein was not entirely convinced of the non-existence of gravitational waves; for a few months after having completed the General Theory, he refocused his efforts on manipulating his equations to obtain an equation that looked like Maxwell's wave equation of electrodynamics, which predicts the existence of electromagnetic waves.

However, these equations are complex, and Einstein had to make several approximations and assumptions to transform them into something similar to Maxwell's equations. For some months, his efforts were futile. The reason was that he used a coordinate system that hindered his calculations. When, at the suggestion of a colleague, he changed coordinate systems, he found a solution that predicted three different kinds of gravitational waves. These three kinds of waves were baptised by Hermann Weyl as longitudinal-longitudinal, transverse-longitudinal, and transverse-transverse.

These approaches made by Einstein were long open to criticism from several researchers, and even Einstein had doubts. Yet the question of the existence of these gravitational waves dogged Einstein and other notable figures in the field of relativity for decades to come. By 1922, English astrophysicist Arthur Eddington wrote an article entitled "The Propagation of Gravitational Waves". According to Eddington's explanation, out of the three types of gravitational waves enunciated by Herman Weyl, only the 'transverse-transverse' wave type propagates at the speed of light in all coordinate systems, so he did not rule out its existence.

In 1933, Einstein emigrated to the United States, where he held a professorship at the Institute for Advanced Study in Princeton. Among other projects, he continued to work on gravitational waves with the young American student Nathan Rosen.

In 1936, Einstein and Rosen sent an article entitled "Are there any gravitational waves?" to the prestigious journal Physical Review on 1 June. Although the original version of the manuscript does not exist today, it is evident that the answer to the article's title was "They do not exist." In the end, Einstein rather inadvertently became convinced of the existence of gravitational waves, whereas Nathan Rosen always thought that they were just a formal mathematical construct with no real physical meaning.

Pirani's Work

In 1956, Felix A E Pirani published a work that became a classic article in the further development of the Theory of Relativity and gravitational waves. The article's title was "On the physical significance of the Riemann tensor". The intention of this work was to demonstrate a mathematical formalism for the deduction of physical observable quantities applicable to gravitational waves. The importance of Pirani's paper is that he used a very practical approach that got around the whole problem of the coordinate system, and he showed that the waves would move particles back and forth as they pass by.

One of the most famous of Einstein's collaborators, Peter Bergmann, wrote a well-known popular book "The Riddle of Gravitation", in which he describes the effect of a gravitational wave passing over a set of particles.

When a gravitational wave passes through a set of particles positioned in an imaginary circle and initially at rest, the passing wave will move these particles. This motion is perpendicular (transverse) to the direction in which the gravitational wave travels. For example, suppose a gravitational pulse passes in a direction perpendicular to this page, the set of particles is disturbed as shown below.

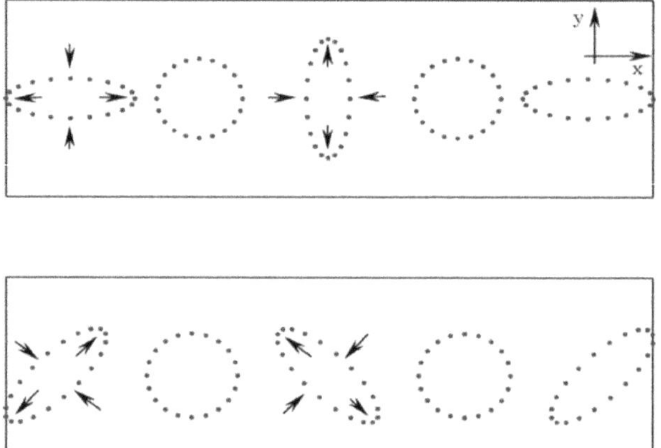

Figure 2-1. *shows the set of particles disturbed (top) in horizontal elliptical mode and (bottom) in inclined elliptical mode.*

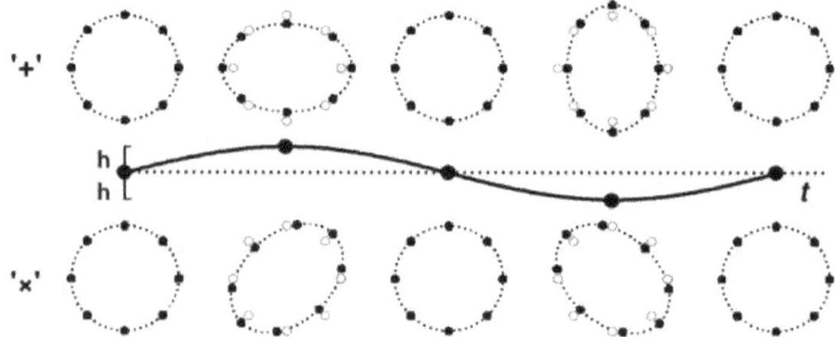

Figure 2-2. *shows how a set of particles, initially arranged in a circle, would sequentially move.*

All these motions occur successively in the plane perpendicular (transverse) to the direction of wave propagation.

With the above idea, it seems the detection of gravitational waves is very simple; on the contrary, it becomes a matter of concern when the law of conservation of energy is visualised for General Relativity.

Chapel Hill Conference

During the mid-1950s, the question of whether or not gravitational waves would transmit energy was still a hot issue. In addition, the controversy could not be solved since there were no experimental observations that would settle this matter. However, this situation was finally clarified thanks to the already mentioned work by Pirani and the comments suggested by Richard Feynman together with a hypothetical experiment he proposed. The experiment was suggested, and comments were delivered by Feynman during a milestone Congress held in 1957 in Chapel Hill, North Carolina. The meeting brought together many scientists and young physicists of the new guard: Feynman, Schwinger, Wheeler, and others. In addition to the issues and debates on the cosmological models and the reality of gravitational waves, during the conference, many questions were formulated, and one of them, of course, is whether gravitational waves carry energy or not.

During the discussions, Feynman came up with an argument that gravitational waves carry energy, which convinced most of the audience.

His reasoning is today known as the "sticky bead argument". Feynman's reasoning is based on a thought experiment that can be described briefly as follows:

Imagine two rings of beads on a bar (see Figure 2-3, upper part). The bead rings can slide freely along the bar. If the bar is placed transversely to the propagation of a gravitational wave, the wave will generate tidal forces with respect to the midpoint of the bar. These forces, in turn, will produce longitudinal compressive stress on the bar.

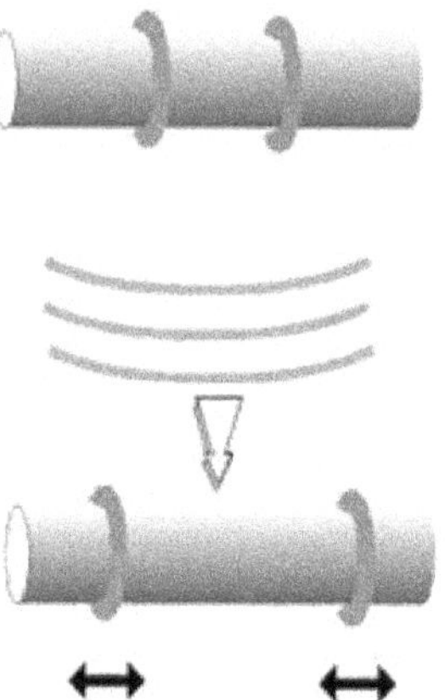

Figure 2-3.   Sketch of the "Sticky Bead Argument".

Meanwhile, because the bead rings can slide on the bar and also in response to the tidal forces, they will slide towards the extreme ends first and then to the centre of the bar (Figure 2-3, bottom). If contact between the beads and the bar is "sticky", then both pieces (beads and bar) will be heated by friction. This heating implies that energy was transmitted to the bar by the gravitational wave, showing that gravitational waves carry energy.

Weber's Experiments

Among the Chapel Hill audience, the maverick American physicist Joseph Weber was present. Weber was an engineer at the University of Maryland. He became fascinated by discussions about gravitational waves and decided to design a device that could detect them. Thus, while discussions among theoretical physicists continued in subsequent years, Weber went even further and soon began designing an instrument to make the discovery.

As a gravitational wave passes by, it first squeezes and then stretches space, which makes objects squeeze and stretch accordingly. An event, such as the collision of two black holes, which produces an intense burst of radiation might briefly stretch a nearby object to twice its original length; a human being 2 metres tall would be stretched to 4 metres, with unpleasant results. But if the same event occurred as far away from us as the centre of the Milky Way, an object one metre long here on Earth would be stretched by a billionth of a billionth (10^{-18}) of a metre, very roughly a hundred-millionth of the size of an atom. Weber published a paper in 1960 setting out how it might be possible to measure such effects and then set out to build an experiment to do just that.

The key to Weber's idea was that although a gravitational wave passing through a solid metal cylinder would stretch and squeeze the cylinder only by a tiny amount, if the cylinder were just the right size this would set it ringing, briefly, like a bell struck with a hammer. Weber's electronic expertise came in handy with the design of the instruments to measure very small vibrations of the bar—a ring of detectors around the waist of the cylinder to convert the vibrations into electric signals that could be recorded and analysed.

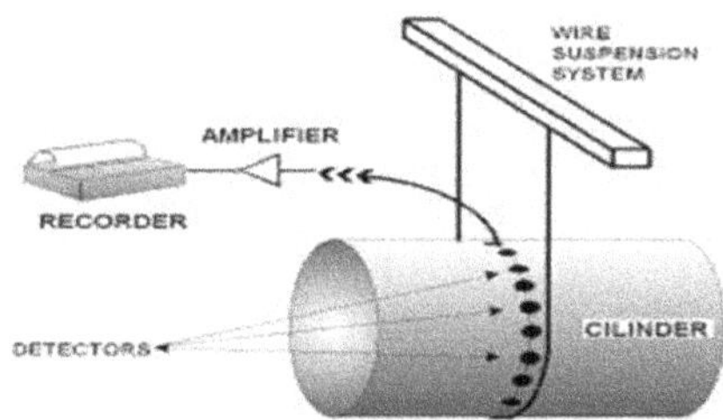

Figure 2-4. Sketch of Weber's cylinder detector and a photo of Joseph Weber at the antenna.

Joseph Weber

Weber's first detector was constructed with the assistance of Robert Forward, a graduate student who went on to become a top physicist and also a leading science fiction writer. It was a cylinder made of aluminium and was 5 feet long.

Weber built several similar bars, all in the same location, but realised that they were being affected by local disturbances such as trucks rolling past. So, he set up two identical bars, each of them 5 feet long and weighing 3,100 pounds, one installed in Maryland and the other 700 miles away at a lab near Chicago. He reasoned that only gravitational waves would affect both bars simultaneously (actually, with a tiny delay, assuming the waves travel at the speed of light), and they were linked by a telephone line to record any 'coincidences'— disturbances that affected both detectors within a 0.44-second time gap or 'window'.

Starting in December 1968, Weber and his team recorded several events that they could not explain by any outside influences. When Weber announced his results at a meeting of 'relativists' in June 1969, they caused a sensation. Many other physicists were motivated to take up the search for gravitational waves, and several of them visited Weber's lab to study his equipment for themselves.

Theoretical Physicists of the time like Dennis W Sciama and Martin John Rees of England, Richard Field and Charles Misner of America, and Peter Kafka of Germany, after their theoretical considerations, were sceptical about Weber's measurements and credibility of his results.

Efforts to follow up Weber's apparent discovery were soon underway in the Soviet Union, Italy, Germany, Japan, and both in England and Scotland. In the United States alone, gravitational wave experiments were developed at Bell Laboratories, the IBM research centre in New York, the University of Rochester, and Stanford University.

As Tony Tyson of Bell Labs said in a scientific paper published in 1972, "It is clear that if it were not for Weber's work, we would not be as near as we are today to the possible detection of gravitational radiation." A variety of detectors were devised with the dual aims of improved sensitivity compared with Weber's bars and the ability to detect different wavelengths of radiation. But nobody found anything, even though Weber continued to report detections of what he thought were gravitational wave events.

Weber never accepted that his detectors were not actually detecting gravitational waves, and the mystery of just what they were detecting, if anything, has never been resolved. He continued working with his bars for the rest of his life, becoming isolated from the mainstream of gravitational wave detector research. His death was indirectly caused by his obsession with gravitational waves. In the winter of early 2000, a few months short of his 81st birthday, he was visiting his unmanned observatory on a hill in Maryland when he slipped on ice and broke several bones. He never recovered from complications caused by the injuries and died 8 months later on 30 September.

He did not live to see the LIGO experiment become operational. In an interview that is now in the archives of the American Institute of Physics, Charles Misner, a leading relativist, said, "The whole effort would never have been started if he hadn't shown the world that you could take gravitational waves seriously." One measure of Weber's place in history is that his first bar is now among the scientific artefacts at the Smithsonian Institution in Washington, DC. Joseph Weber is considered a pioneer in experimental gravitation and therefore he is honoured by the American Astronomical Society, which awards the Joseph Weber Award for Astronomical Instrumentation every year.

However, the invalidation of Weber's results urged other researchers to redouble the search for gravitational waves or devise indirect methods of detection. Weber's results stimulated other experimental groups to start their own experiments. By 1975, almost all people in the gravitational wave (GW) detection community agreed that Weber had not detected gravitational waves. Nevertheless, the search for gravitational waves continued with even more intensity.

By the mid-seventies, several detectors were already operative and offered many improvements over Weber's original design; some cylinders were even cooled to reduce thermal noise. These experiments were operating in several places: at Bell Labs Rochester-Holmdel; the University of Glasgow, Scotland; an Italo-German joint programme in Munich and Frascati; Moscow; Tokyo; and the IBM labs in Yorktown Heights. As soon as these new instruments were put into operation, a common pattern emerged: there were no signals. In the late seventies, everyone except Weber himself agreed that his proclaimed detections were spurious. At that time, great pessimism and disappointment reigned among the "seekers" of gravitational waves.

Hulse-Taylor Pulsar

However, in 1974, an event occurred that raised hopes. In that year Joseph Hooten Taylor and Alan Russell Hulse of the University of Massachusetts Amherst, United States found an object in the sky—a binary pulsar (pulsating radio star)—that revealed that an accelerated mass radiated gravitational energy. They were carrying out a pulsar survey at the Arecibo Observatory in Puerto Rico, a radio telescope with a 305-metre (1000 ft) dome. You may remember Arecibo as that big telescope that collapsed to rubble in late 2020 due to lack of funding and neglect. There's an important reason why Arecibo was so big. The goal of radio telescopes is to detect radio waves whose wavelengths are often more than the Earth's radius. The sources of radio waves outside the solar system are really weak, so we need very big dishes to detect those objects, and Arecibo successfully detected something.

Figure 2-5.   The Arecibo Observatory in Puerto Rico. Credit: NAIC Arecibo Observatory, a facility of the NSF.

The scientists detected a "weird" pulsar, later named PSR B1913+16 or the "Hulse-Taylor binary". The 'PSR' is shorthand for pulsar, and the numbers indicate the position of the object in the sky, like latitude and longitude on Earth. It was a particularly exciting discovery because the beam from the pulsar was sweeping round so fast that it produced 17 blips every second, meaning that the neutron star was spinning once every 58.98 milliseconds, making it the second-fastest pulsar known at the time. But Hulse meticulously observed and found that in an hour, the period of this particular pulsar had changed by 27 milliseconds, which was an unheard-of "error" for a pulsar. At first, Hulse thought that he was going wrong somewhere. But measurement after measurement showed that the pulsar's period really was changing. Sometimes it increased by a few milliseconds, sometimes it decreased. Hardly able to believe what he had found, Hulse realised that the changes were caused by the pulsar orbiting around another star—making it a member of a binary pair, with each star orbiting around their common centre of mass. When the pulsar is moving towards the Earth, the pulses are piled up closer to one another; when it is moving away, the gap between them is stretched out. This is a variation expected of the well-known Doppler effect. The speed with which the changes are occurring showed that the object that PSR 1913+16 is orbiting around must be another neutron star.

Together, Hulse and Taylor determined the "orbital parameters" of the system. The pulsar zips around its companion once every 7 hours 45 minutes, reaching a maximum speed of 300 km per second (a thousandth of the speed of light), with an average speed of 200 km per second. Then astronomers immediately realised that such a system could act as a test bed for General Relativity, and in particular that it ought to be producing gravitational waves. Putting together all the data collected and the measurements of the way the "ticking" of the pulsar changes over an orbit could be used to determine the ratio of the masses of the two stars. It was found that the pulsar has a mass 1.44 times that of the Sun, while its companion has a mass just under 1.39 times the mass of the Sun.

According to the General Relativity theory, two neutron stars separated by less than the diameter of the Sun orbit around one another in an inspiral should lose energy in the form of emission of gravitational waves. The prediction for the binary pulsar was that the orbital period, which is about 27,000 seconds, would decrease by about 0.0000003 per cent, or 75 millionths of a second, each year. After making allowances for all kinds of disturbances and analysing some 5 million pulses from PSR 1913+16 over a period of 3 years, in December 1978, Taylor was able to announce the result to an international meeting in Germany. The orbit of the binary pulsar was "decaying" exactly in line with the predictions of the General Relativity theory. This was spectacular confirmation that the general theory is right, and gravitational waves are real. In 1993, Hulse and Taylor shared the Nobel Prize in physics for their discovery.

Joseph Hootoon Taylor   **Alan Russell Hulse**

Pulsars

A pulsar is a highly magnetised spinning neutron star that emits beams of electromagnetic radiation out of its magnetic poles. A neutron star is formed when the core of a violently exploding star, called a supernova, collapses and compresses together. Neutron stars are the second-most compact objects in our Universe after black holes. In fact, a teaspoon of neutron star mass weighs a staggering hundred million tonnes. Its diameter is about 20 km or less, but masses range between 1.18 and 1.97 times that of the Sun. Neutrons, at the surface of the star, decay into protons and electrons. As these charged particles are released from the surface, they enter an intense magnetic field that surrounds the star and rotates along with it. Accelerated to speeds approaching that of light, the particles give off electromagnetic radiation by synchrotron emission. This radiation is released as intense beams from the pulsar's magnetic poles.

A pulsar's magnetic field is on the order of 10^{12} Gauss while the magnetic field of Earth is only 0.5 Gauss. Rotating at a rate of up to hundreds of times per second, a pulsar emits a beam of particles and light, including strong radio waves, out of each of its magnetic poles with the magnetic axis usually "precessing" around the rotation axis so that each beam sweeps out a conical path in the sky. If a pulsar is oriented such that the solar system lies on this conical path, we can use radio telescopes to detect a "blip" of signal each time its beam comes our way. In fact, this regular signal, which we see at a number of different wavelengths, is our only evidence that pulsars exist.

A pulsar is a special kind of neutron star that blasts out two beams of radio waves in opposite directions. As the dead star spins, those beams sweep through space like the lights on a lighthouse. If Earth is in the path of one of those beams, we see a flash of radio waves every time it sweeps past us. That makes the pulsar appear to pulse at very regular intervals.

Some of the fastest pulsars that rotate once every few milliseconds are particularly useful tools because the arrival times of the pulses at our telescopes are so reliably regular that they can be used as extremely precise clocks, even rivalling some atomic clocks.

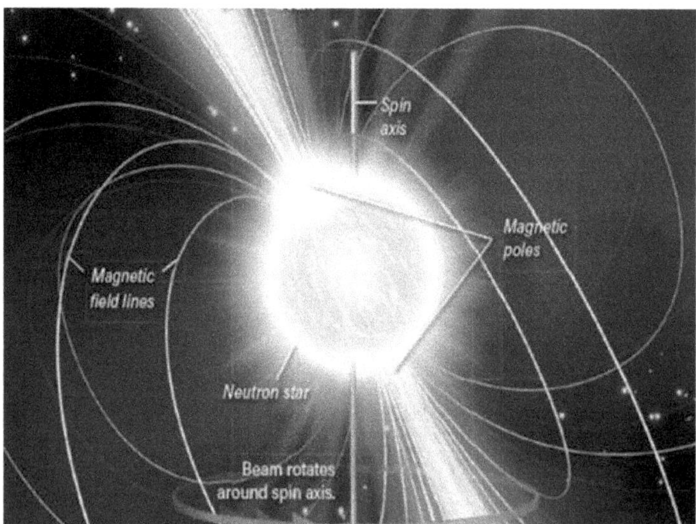

Figure 2-6. This picture shows how an artist imagines a pulsar. The dead star blasts out two beams of radio waves. As the pulsar spins, the radio waves sweep through space like lights from a lighthouse.

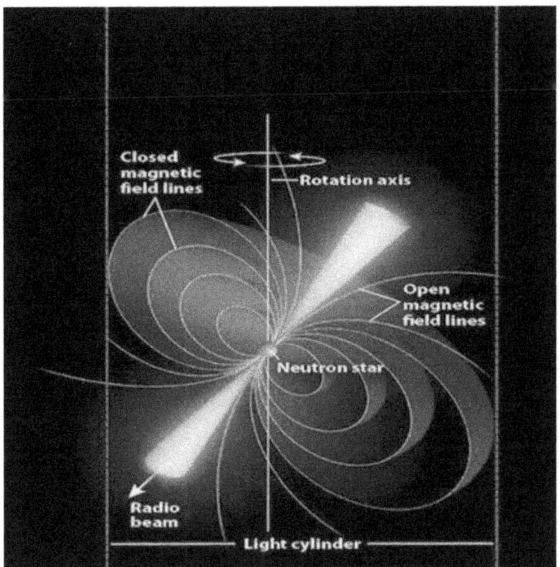

Figure 2-7.   Schematic of a Pulsar

Antony Hewish and Jocelyn Bell Burnell, astronomers working at Cambridge University, first discovered pulsars in 1967 with the aid of a radio telescope specially designed to record rapid fluctuations in radio sources. Initially, some scientists thought the radio beams might be coming from aliens because the pulses were so regular. However, Bell Burnell found radio pulses coming from a different part of space, far from the first signal. It was unlikely that two groups of aliens were signalling us at the same time from so far apart, so scientists looked for a different explanation. They eventually learned that the radio waves were coming from pulsars scattered throughout space.

If a gravitational wave passes between the pulsar and us, it would alternately stretch and compress the distance that a pulse of light from the pulsar has to travel before it reaches our telescopes. As light moves at a constant speed, the pulses would take a longer or shorter amount of time, respectively, to travel to us resulting in those blips arriving later or earlier than if there were no gravitational waves at all.

Joseph Taylor and Russell Hulse won the Nobel Prize for Physics in 1993 for their study of timing variations in the pulsar PSR 1913+16. PSR 1913+16 has a companion neutron star with which it is locked in a tight orbit. The two stars' enormous interacting gravitational fields affect the regularity of the radio pulses, and by timing these and analysing their variations, Taylor and Hulse found that the stars were rotating ever faster around each other in an increasingly tight orbit.

Binary Pulsar PSR B1913+16

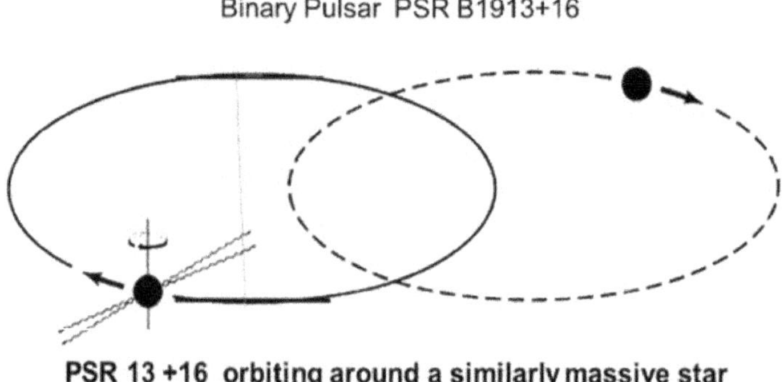

Figure 2-8.   Binary Pulsar PSR B1913+16

The two neutron stars in the binary pulsar are 1.438 and 1.390 times the mass of our Sun. The binary pulsar PSR B1913+16 is 21 thousand light-years from Earth. The elliptical orbit has an eccentricity equal to 0.617 and a semi-major axis equal to 1.95 million km. The neutron stars complete a rotation in 7 hours and 45 minutes. The rotation distorts spacetime in the vicinity, and energy is radiated out as gravitational waves. Einstein's General Theory of Relativity predicts the exact amount of GW radiation that would be emitted. The radiated energy is lost from the rotational energy of the binary pulsar system, and the neutron stars get progressively closer. The observed change in orbit parameters (the semi-major axis is decreasing by 3.5 metres per year) provides the point of comparison with the prediction of the General Theory. It is estimated that it will take about 300 million years before the neutron stars merge.

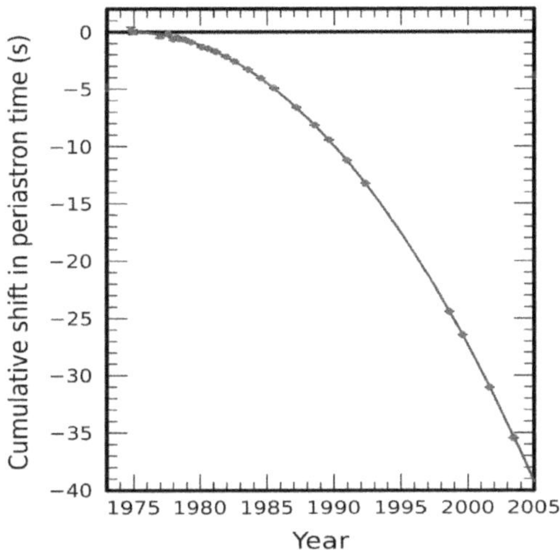

Figure 2-9. Evidence of orbital decay in PSR B1913+16. The data points indicate the observed change in the time of periastron (the point at which an object is closest to the centre of mass of the star it's orbiting) with date, relative to a system not undergoing decay. The parabola illustrates the theoretically expected change according to general relativity.

General Relativity predicts that when two compact stars orbit a common centre of mass, the gravitational waves would carry away the orbital energy and would cause the two stars to draw closer, and eventually merge with each other. This was the first experimental evidence for the existence of the gravitational waves predicted by Albert Einstein in his General Theory of Relativity. While this observation did not directly detect gravitational waves, it pointed to their existence. These studies on pulsars sparked renewed interest in the future discovery of gravitational waves and urged other researchers to redouble the search for the lost waves and devise other methods of detection.

Entry of Interferometers

Let us now turn our attention to the origin of the interferometer method. The first explicit suggestion of a Laser Interferometer detector was outlined in the former USSR by Gertsenshtein and Pustovoid in 1962. The idea was not carried out and eventually was resurrected in 1966 behind the "Iron Curtain" by Vladimir B Braginski, but then again fell into oblivion due to a lack of funds.

In the early seventies, Robert L Forward, a former student of Joseph Weber at that time working for Hughes Research Laboratory in Malibu, California, decided, with the encouragement of Rainer Weiss, to build a laboratory interferometer with Hughes' funds. Forward's interest in interferometer detectors had evolved some years before when he worked for Joseph Weber at his laboratory at the University of Maryland on the development and construction of Weber's antenna. After 150 hours of observation with his 8.5-m arms interferometer, Robert Forward reported "an absence of significant correlation between the interferometer and several Weber bars detectors, operating at Maryland, Argonne, Glasgow, and Frascati." In short, Robert Forward did not observe gravitational waves.

Also in the 1970s, Reiner Weiss at MIT independently conceived the idea of building a Laser Interferometer inspired by an article written by Felix Pirani. After a while, Weiss started building a 1.5m long interferometer prototype in the RLE (Research Laboratory of Electronics) at MIT using military funds. Sometime later, a law was enacted in the United States (the "Mansfield Amendment") which prohibited Armed Forces financing projects that were not of strictly military utility. Funding was suddenly suspended. This forced Weiss to seek financing from other US government and private agencies.

In 1974, NSF (National Science Foundation) asked Peter Kafka of the Max Planck Institute in Garching, near Munich, to review a project. The project was submitted by Weiss, who requested \$53,000 in funds for enlarging the construction of a prototype interferometer with arms 9 meters in length. Kafka agreed to review the proposal. Being a theoretician himself, Kafka showed the proposal documents to some experimental physicists at his institute for advice. To Kafka's embarrassment, the local group currently working on Weber bars became very enthusiastic about Weiss's project and decided to build their own prototype, headed by Heinz Billing. The Garching group made the decision to try the interferometer idea. The Germans contacted Weiss for advice, and they also offered a job to one of his students on the condition that he be trained on Weiss's 1.5m prototype. Eventually, Weiss sent David Shoemaker, who had worked on the MIT prototype, to join the Garching group. Shoemaker later helped to build a German 3m prototype and later a 30m interferometer. This interferometer in Garching served for the development of noise suppression methods that would later be used by the LIGO project. It is interesting to mention that Weiss's proposal may seem modest now (9m arms), but he already had in mind large-scale interferometers. NSF then supported Weiss's project, and funds were granted in May 1975.

Around that time, Ronald Drever, then at the University of Glasgow, attended a lecture by Peter Kafka in Munich in March 1975, where the latter highlighted the shortcomings of Weber's experiments in detecting gravitational waves and mooted the idea of laser interferometry methods to study gravitational waves.

After returning to Glasgow, Ronald Drever started developing interferometric techniques and by the end of the 1970s, he was leading a team at Glasgow that had completed a 10m interferometer. Likewise, in 1975, the German group in Munich (Winkler, Rüdiger, Schilling, Schnupp, and Maischberger), under the leadership of Heinz Billing, built a prototype with an arm length of 3m.

The group worked hard to overcome some experimental deficiencies by developing innovative technologies that modern-day gravitational interferometers embraced. In 1983, the same group, now at the Max Planck Institute of Quantum Optics (MPQ) in Garching, improved their first prototype by building a 30m arm-length instrument.

After a couple of years of operating the 30m model, the Garching group was prepared to go for Big Science. In effect, in June 1985, they designed a detailed proposal for a full-sized interferometer of a 3 km length.

The project was submitted for funding to the German authorities, but there was not sufficient interest in Germany at that time, so it was not approved.

In the meantime, similar research was undertaken by the group at Glasgow, now under Jim Hough after Drever's exodus to Caltech. Following the construction of their 10m interferometer, the Scots decided in 1986 to take a further step by designing a Long Baseline Gravitational Wave Observatory. Funds were requested, but their call fell on deaf ears.

Three years later, the Glasgow and the Garching groups decided to unite efforts to collaborate on a plan to build a large detector. It did not take long for both groups to jointly submit a plan for an underground 3-km installation to be constructed in the Harz Mountains in Germany, but again their proposal was not funded. Despite this disheartening ruling, the new partners decided to try for a shorter detector and compensate by employing more advanced and clever techniques. A step forward was finally taken in 1994 when the University of Hanover and the State of Lower Saxony donated land to build a 600m instrument in Ruthe, 20 km south of Hanover. Funding was provided by several German and British agencies.

The construction of GEO 600 (the Gravitational Wave Detector) started on 4 September 1995. The following years of continuous hard work by the British and Germans brought results. Since 2002, the detector has been operated by the Centre for Gravitational Physics, of which the Max Planck Institute is a member, together with Leibniz University in Hanover and Glasgow and Cardiff Universities. The first scientific data run, again together with the LIGO detectors, was performed in August and September 2002. In November 2005, LIGO and GEO instruments began an extended joint science run. In addition to being an excellent observatory, the GEO 600 facility has served as a development and test laboratory for technologies that have been incorporated into other detectors all over the world.

The Origin of the LIGO Project

In the summer of 1975, Weiss went to Dulles Airport in Washington, DC, to pick up Kip Thorne, a renowned theoretical physicist from Caltech, working on the studies of gravitational waves. Kip Thorne had been a student of John Wheeler and took over his mantle, becoming an expert on the General Theory of Relativity and the leading black hole theorist of his generation. The reason for visiting Washington was to attend a NASA meeting on the uses of space research in the field of cosmology and relativity. Weiss recalls, "I picked Kip up at the airport on a hot summer night when Washington, DC, was filled with tourists. He did not have a hotel reservation, so we shared a room for the night."

They did not sleep that night because both spent the night discussing many topics, one among them being how to search for gravitational waves. Thorne decided that night that the first thing they ought to do at Caltech was to construct an interferometric gravitational wave detector. However, he would need help from an experimental physicist. Weiss suggested a name, Ronald Drever. Weiss had only known Drever from his papers, not in person. Drever was famous for the Hughes-Drever Experiments and had also been the leader of the group that built a "Weber cylinder" at the University of Glasgow. At that time, Drever was planning the construction of an interferometer. In 1978, Thorne offered a job to Drever at Caltech for the construction of an interferometer. Drever accepted the offer in 1979, dividing his time between the Scottish university and Caltech. Hiring Drever half-time soon paid dividends because, in 1983, he had already built his first instrument at Caltech, an interferometer whose "arms" measured 40 m.

In 1983 Drever began full-time work at Caltech with the idea of gradually improving and increasing the size of the prototype as it was built and run. In contrast to the Caltech apparatus, as already mentioned, Weiss had built a modest 1.5m prototype with a much smaller budget than the Californian instrument at MIT. In late 1979, the NSF granted modest funds to the Caltech interferometer group and gave a much smaller amount of money to the MIT team. Soon Drever and Weiss began to compete to build more sensitive and sophisticated interferometers.

The sensitivity of the interferometers can be enhanced by boosting the power of the lasers and increasing the optical path of the light beam as it travels through the interferometer arms. To increase the sensitivity of the interferometer, Weiss put forward the use of an optical delay line. In the optical delay method, the laser light passes through a small hole in an adjacent wave divider mirror, and the beam is reflected several times before emerging through the inlet port.

In the meantime, Drever developed an arrangement that utilised "Fabry-Perot cavities". In this method, the light passes through a partially transmitting mirror to enter a resonant cavity flanked at the opposite end by a fully reflecting mirror. Subsequently, the light escapes through the first mirror.

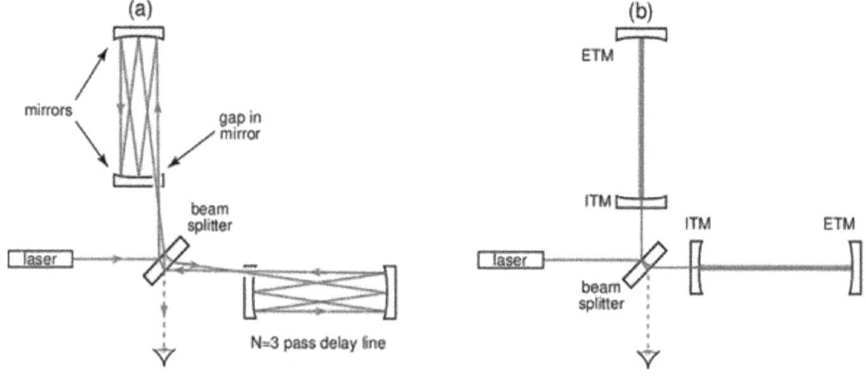

Figure 2-10. Michelson interferometers with (a) delay lines and (b) Fabry-Perot cavities in the arms of the interferometer.

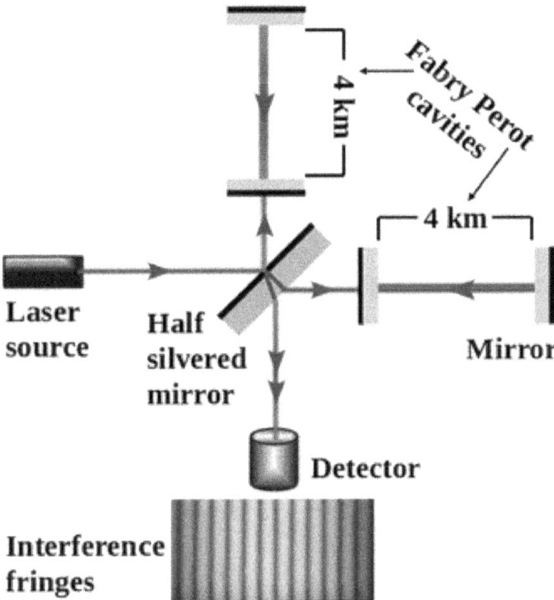

Figure 2-11. The implementation of signal recycling on a Michelson interferometer with Fabry-Perot cavities.

As already mentioned, Weiss experimented with an interferometer whose two L-shaped "arms" were 1.5m long. Drever, meanwhile, had already built and operated a 40m interferometer. With the Caltech Group appearing to be taking the lead, Weiss decided in 1979 to "do something dramatic." Weiss decided to conduct a study in collaboration with industry partners to determine the feasibility and cost of an interferometer whose arms should measure in kilometres.

Weiss, along with his colleagues, worked for 3 years and prepared a document entitled "A Study of a Long Baseline Gravitational Wave Antenna System," co-authored by Peter Saulson and Paul Linsay. This fundamental document is nowadays known as "The Blue Book" and covers many important issues in the construction and operation of such a large interferometer. The Blue Book was submitted to the NSF in October 1983. The proposed budget was just under $100 million to build two instruments located in the United States.

Before "The Blue Book" was submitted for consideration to the NSF, Weiss met with Thorne and Drever at a Relativity Congress in Italy. There they discussed how they could work together; this was mandatory because the NSF would not fund two megaprojects on the same subject and with the same objective. However, from the very beginning, it was clear that Drever did not want to collaborate with Weiss, and Thorne had to act as a mediator. In fact, the NSF settled matters by integrating the MIT and Caltech groups together in a "shotgun wedding" so the "Caltech–MIT" project could be jointly submitted to the NSF.

The Caltech–MIT project was funded by the NSF and named the "Laser Interferometer Gravitational-Wave Observatory", known by its acronym LIGO. The project was led by a triumvirate of Thorne, Weiss, and Drever. Soon, the interactions between Drever and Weiss became difficult because, besides the strenuous nature of their interaction, both had differing opinions on technical issues.

During the years 1984 and 1985, the LIGO project suffered many delays due to multiple discussions between Drever and Weiss, mediated, when possible, by Thorne. In 1986, the NSF called for the dissolution of the triumvirate of Thorne, Drever, and Weiss. Instead, Rochus E Vogt was appointed as a single project manager.

In 1988, the project was finally funded by the NSF. From that date until the early 1990s, the project progress was slow and underwent a restructuring in 1992. As a result, Drever stopped being part of the project, and in 1994, Vogt was replaced by a new director, Barry Clark Barish, an experimental physicist who was an expert in high-energy physics. Barish had experience in managing big projects in physics. His first activity was to review and substantially amend the original 5-year-old NSF proposal. With its new administrative leadership, the project received good financial support. Barish's plan was to build the LIGO as an evolutionary laboratory where the first stage, "initial LIGO" (or iLIGO), would aim to test the concept and offer the possibility of detecting gravitational waves. In the second stage ("aLIGO" or Advanced LIGO), wave detection would be very likely.

Two observatories, one in Hanford in Washington State and another in Livingston, Louisiana, were built. Construction began in late 1994 and early 1995, respectively, and ended in 1997. Barish's idea to make LIGO an evolutionary apparatus proved in the end to pay dividends. The idea was to produce an installation whose parts (vacuum system, optics, suspension systems, etc) could always be readily improved and buildings that could house those ever-improving interferometer components.

In effect, LIGO was incrementally improved by advances made in its own laboratories and those due to associations with other laboratories (VIRGO and GEO600). To name some: Signal recycling mirrors were first used in the GEO600 detector, as well as the monolithic fibre-optic suspension system that was introduced into Advanced LIGO. In brief, LIGO's detection was the result of a worldwide collaboration that helped it evolve into its present remarkably sensitive state.

The initial LIGO, operated between 2002 and 2010, did not detect gravitational waves. The upgrade of LIGO (Advanced LIGO) began in 2010 to replace the detection and noise suppression and improve stability operations at both facilities. This upgrade took 5 years and had contributions from many sources.

Advanced LIGO (aLIGO) started its operations in February 2015. The team operated in "engineering mode"—that is, in test mode—and in late September began scientific observation. It did not take many days for LIGO to detect gravitational waves. The mystery of gravitational waves is finally unfolded. The rest is history.

CHAPTER – 3

Sources of Gravitational Waves

Technically speaking, every massive object that accelerates produces gravitational waves. This includes humans, cars, aeroplanes, etc, but the masses and accelerations of objects on Earth are far too small to make gravitational waves big enough to detect with our instruments. To find big enough gravitational waves, we have to look far outside of our own solar system.

In general, any accelerating body that is not spherically or cylindrically symmetric will produce a gravitational wave. Consider a star that goes into a supernova. This explosion will produce gravitational waves if the mass is not ejected in a spherically symmetric way, although the centre of mass may be in the same position before and after the explosion. Another example is a spinning star. A perfectly spherical star will not produce a gravitational wave, but a lumpy star will.

The gravitational waves that modern detectors are sensitive to would be in the audible frequency range if they were sound waves. In that sense, these detectors can be thought of as "gravitational wave radios". Just like radio waves cannot be heard without a radio to detect the radio waves and decode the music signal to send to the speakers, gravitational waves too cannot be heard without a detector to distinguish the gravitational wave and send that signal to speakers. All of the physics that went into the production of a gravitational wave is then encoded in this "music" for physicists to decode.

What Can Generate Gravitational Radiation?

As we are going to learn (in the next chapter) that gravitational radiation must be transverse and traceless, we can draw some general conclusions about the type of motion that can generate such radiation. A perfectly spherically symmetric variation is only monopolar and does not change the trace-free quadrupole, so it does not produce radiation. No matter how violent an explosion or a collapse (even into a black hole!), no gravitational radiation is emitted if spherical symmetry is maintained.

In addition, a rotation that preserves axisymmetry (without contraction or expansion) does not generate gravitational radiation because the quadrupolar and higher moments are unaltered. For example, a neutron star can rotate rigidly, arbitrarily, and rapidly without emitting gravitational radiation as long as it maintains axisymmetry about its rotation axis.

This immediately allows us to focus on the most promising types of sources for gravitational-wave emission. The general categories are binary systems, continuous wave sources (rotating stars with non-axisymmetric lumps), bursts of radiation (e.g., asymmetric collapses), and stochastic sources (i.e., individually unresolved sources with random phases); arguably the most interesting of these would be a background of gravitational waves from the early Universe.

Types of Gravitational Waves

The Universe is filled with incredibly massive objects undergoing rapid accelerations that generate gravitational waves that we can now detect. Known objects are pairs of black holes or neutron stars orbiting each other, a neutron star and black hole orbiting each other, or gigantic stars blowing themselves up at the ends of their lives. Broadly speaking, there are four main sources of gravitational waves caused by different kinds of motion and changing distributions of mass: continuous, inspiral, stochastic, and burst gravitational waves.

I. CONTINUOUS GRAVITATIONAL WAVES

Continuous gravitational waves are produced by systems that have a fairly constant and well-defined frequency. Examples of these are binary systems of stars or black holes orbiting each other (long before merger), or a single star swiftly rotating about its axis with a large mountain or other irregularity on it. These sources are expected to produce comparatively weak gravitational waves since they evolve over longer periods of time and are usually less catastrophic than sources producing inspiral or burst gravitational waves. The sound these gravitational waves would produce is a continuous tone since their frequency is nearly constant.

The most common sources of continuous gravitational waves are rotating asymmetric astronomical objects, such as:

1. Neutron Stars: Neutron stars (NS) are the collapsed cores of supermassive giant stars that contain between 10 and 25 solar masses. Aside from black holes, they are the densest objects in the Universe. Neutron stars are extremely dense remnants of massive stars that have undergone a supernova explosion. Their journey from a main sequence star to a collapsed stellar remnant is a fascinating scientific story. If a neutron star is not perfectly spherical and has some deformity or "mountain" on its surface, it can emit continuous gravitational waves as it rotates.

2. Pulsars: Pulsars are highly magnetised, rapidly rotating neutron stars that emit beams of electromagnetic radiation. If these beams are not perfectly aligned with the rotational axis, they produce continuous gravitational waves.

3. Non-axisymmetric Supernovae: Some supernovae may not result in perfectly symmetric explosions, leaving behind remnants that emit continuous gravitational waves as they rotate.

SUPERNOVA

A supernova is the explosion of a star at the end of its life. It can emit more energy in a few seconds than our Sun will radiate in its lifetime of billions of years. A supernova is the largest explosion that takes place in space. Scientists have identified several types of supernovae. Understanding these different types of supernovae is not just a matter of cosmic classification; it provides crucial insights into the life and death of stars.

- Type I supernovae: No hydrogen.

- Type II supernovae: Shows hydrogen.

- Type III supernovae: Electron-capture.

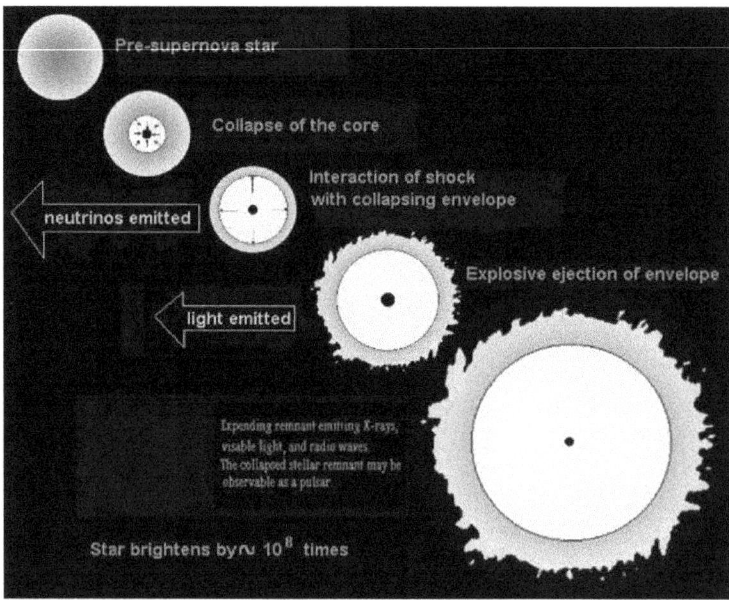

Figure 3-1.   Supernova collapse sequence

Within about 0.1 seconds, the core collapses and gravitational waves are emitted. After about 0.5 seconds, the collapsing envelope interacts with the outward shock. Neutrinos are emitted. Within two hours, the envelope of the star is explosively ejected. When the photons reach the surface of the star, they brighten by a factor of 100 million. Over a period of months, the expanding remnant emits X-rays, visible light, and radio waves in a decreasing fashion.

Continuous gravitational waves have a stable frequency, and their detection requires long-duration observations to accurately identify their signal amidst background noise. Unlike transient signals, which are typically shorter in duration but stronger in amplitude, continuous gravitational waves are weaker and may require sophisticated data analysis techniques to tease out their presence from instrumental and environmental noise.

Various ground-based gravitational wave observatories, such as the Laser Interferometer Gravitational-Wave Observatory (LIGO) and Virgo, are equipped to search for continuous gravitational waves. The continuous wave searches involve extensive data analysis over extended periods to identify any subtle, persistent signals from rotating astrophysical sources. Detecting continuous gravitational waves can provide valuable information about the properties of rotating astronomical objects and the extreme physics in their vicinity.

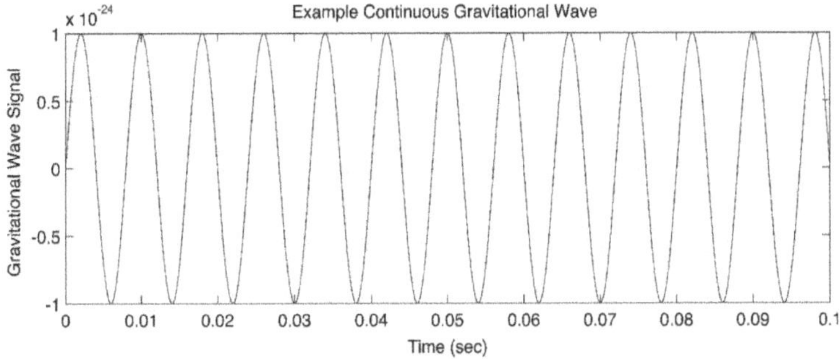

Figure 3-2. *A short snippet of an example gravitational-wave strain signal from a continuous gravitational-wave source. Note how the peaks and troughs of the waves come at equal time intervals (hence, constant frequency) and how the height of each peak and trough is the same (constant amplitude). [Image: A. Stuver/ LIGO]*

II. INSPIRAL GRAVITATIONAL WAVES

Figure 3-3.   An artist's impression of two stars orbiting each other and progressing (from left to right) towards a merger, resulting in gravitational waves. [Image: NASA/CXC/GSFC/T.Strohmayer]

Inspiral gravitational waves are generated during the end-of-life stage of binary systems where the two objects merge into one. These systems are usually two neutron stars, two black holes, or a neutron star and a black hole whose orbits have degraded to the point that the two masses are about to coalesce. As the two masses rotate around each other, their orbital distances decrease and their speeds increase, much like a spinning figure skater who draws his or her arms in close to their body. This causes the frequency of the gravitational waves to increase until the moment of coalescence.

The sound these gravitational waves would produce is a chirping sound (much like when increasing the pitch rapidly on a slide whistle) since the binary system's orbital frequency is increasing (any increase in frequency corresponds to an increase in pitch).

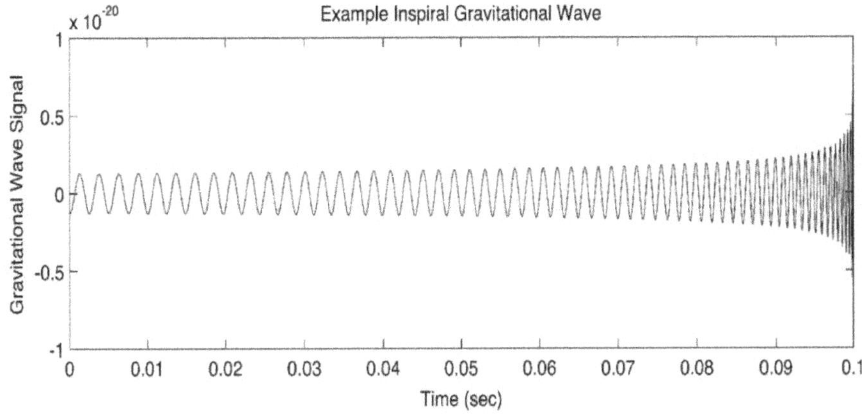

Figure 3-4.   Inspiral gravitational waves

III. STOCHASTIC gravitational waves

Stochastic gravitational waves refer to a background of gravitational waves that are random in nature and do not have a specific identifiable source. The physical processes occurring at the earliest moments of the Universe certainly created a stochastic background that exists, at some level, today. This is analogous to the cosmic microwave background (CMB), which is an electromagnetic record of the early Universe and is likely to be the leftover light from the Big Bang.

However, we do presume that many small gravitational waves are passing by from all over the Universe all the time and that they are mixed together at random. These small waves from every direction make up what is called a "Stochastic Signal", so called because the word 'stochastic' means having a random pattern that may be analysed statistically but not predicted precisely. These will be the smallest and most difficult gravitational waves to detect, but it is possible that at least part of this stochastic signal may originate from the Big Bang.

If these stochastic gravitational waves truly originated in the Big Bang, they will have been stretched as the Universe expanded and they can tell us about the true nature of the very beginning of the Universe.

They would have been produced between approximately 10^{-36} to 10^{-32} seconds after the Big Bang, whereas the CMB was produced approximately 300,000 years after the Big Bang.

Unlike gravitational waves produced by distinct astrophysical events, such as the collision of black holes or neutron stars, stochastic gravitational waves arise from a superposition of many unresolved and possibly unrelated sources distributed throughout the Universe. These sources could be of both cosmological and astrophysical origins. Processes in the early Universe should have produced a stochastic background of gravitational waves that extend through the entire frequency range from extremely low frequencies $f \sim 10^{-18}$ Hz to the high-frequency band $f \sim 10^4$ Hz and beyond.

There are several potential sources that contribute to the stochastic gravitational wave background, including:

1. Cosmological Phase Transitions: During the early Universe, the cosmos might have undergone phase transitions, leading to fluctuations in spacetime that generate gravitational waves.

2. Primordial Black Holes: Small black holes formed in the early Universe; they could emit gravitational waves as they interact or merge with each other.

3. Cosmic Strings: These are theoretical one-dimensional topological defects that could form in the early Universe. As cosmic strings move and interact, they produce gravitational waves.

4. Inflation: The rapid expansion of the Universe during the inflationary epoch could have generated gravitational waves, leaving behind a stochastic background.

5. Unknown Astrophysical Sources: There may be other unresolved sources of gravitational waves from various astrophysical phenomena that contribute to the stochastic background.

6.   Detecting stochastic gravitational waves is challenging because they create a nearly isotropic and homogeneous signal across the sky. It requires precise measurements over extended periods to distinguish this background from instrumental noise and other astrophysical sources.

Efforts to detect stochastic gravitational waves are ongoing and involve collaborations between various ground-based and space-based gravitational wave observatories. These observations can provide valuable insights into the early Universe, the nature of gravity, and the distribution of unresolved sources throughout the cosmos.

The recent observations of gravitational waves by the Advanced LIGO and Advanced Virgo detectors imply that there is also a stochastic background that has been created by binary black holes and binary neutron star mergers throughout the history of the Universe.

Detecting relic gravitational waves from the Big Bang will allow us to see further back into the history of the Universe than ever before.

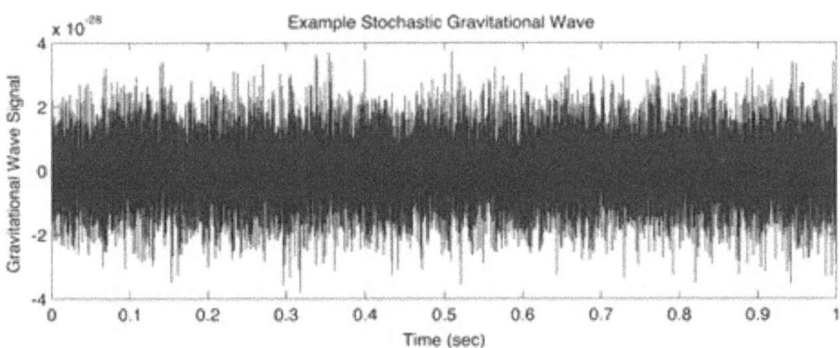

Figure 3-5.   Schematic of Stochastic gravitational waves

IV. Burst gravitational waves

The search for "burst gravitational waves" is truly a search for the unexpected because LIGO has yet to detect them, and also because there are still so many unknowns that we really don't know what to expect.

For example, sometimes we don't know enough about the physics of a system to predict how gravitational waves from that source will appear.

We also expect to detect gravitational waves from systems we never knew about before. To search for these kinds of gravitational waves, we cannot assume that they will have well-defined properties like those of continuous and compact binary inspiral waves. This means we cannot restrict our analyses to searching only for the signatures of gravitational waves that scientists have predicted.

There are hypotheses that some systems such as supernovae or gamma-ray bursts may produce burst gravitational waves.

Astrophysical sources: – Core collapse supernova. – Neutron star instabilities. – Fallback accretion onto a neutron star. – Non-axisymmetric deformation in magnetars. – Pulsar glitches. – Neutron star post-mergers. – Black hole accretion disk fragmentation. –...

- Cosmological sources: Cosmic strings

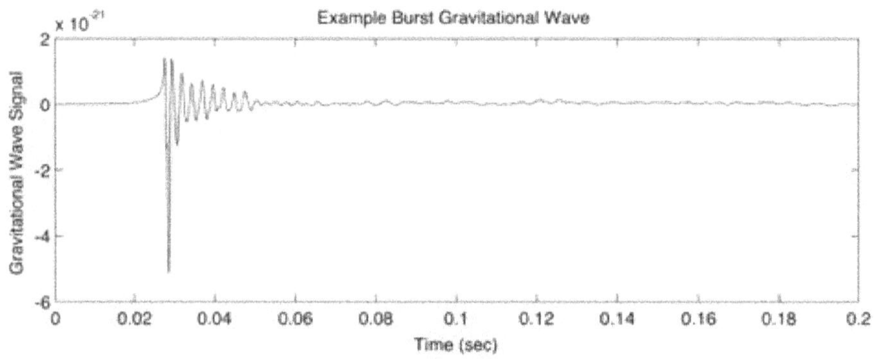

Figure 3-6. An example of signals from a burst gravitational wave source. (Image: A. Stuver/LIGO using data from C. Ott, D. Burrows, et al)

Searching for burst gravitational waves requires being utterly open-minded. For these kinds of gravitational waves, scientists must recognise a pattern of signals even when such a pattern has not been modelled before. If we don't know what we're looking for, it's really hard to find it. While this makes searching for burst gravitational waves difficult, detecting them has the greatest potential to reveal revolutionary information about the Universe.

CHAPTER – 4

Properties of Gravitational Waves

A gravitational wave is a travelling gravitational field. An electromagnetic wave is a combination of a travelling electric field and a magnetic field, both transverse to the direction of propagation. Similarly, the effects of a gravitational wave are transverse to the direction of propagation and are similar to a tidal gravitational field. In terms of General Relativity, a gravitational wave will stretch one dimension of space while contracting the other.

Like electromagnetic waves, gravitational waves are characterised by a wavelength λ and frequency f that travel with the speed of light c given by $c = f \lambda$ and carry energy and momentum with them. Gravitational waves are expected to have frequencies over a wide range in the order of 10^{-16} Hz $< f < 10^4$ Hz.

The problem with gravitational waves is that their effect on matter is almost negligible. Furthermore, not all gravitational waves are equal, as this depends on the phenomenon that generates them (as explained in chapter 3). To evaluate the intensity of the effect that a particular wave produces on the matter, a dimensionless factor, denoted by the letter "h", has been introduced. The dimensionless amplitude "h" describes the maximum displacement per unit length that would produce waves on an object. To illustrate the definition of "h", refer to the following figure:

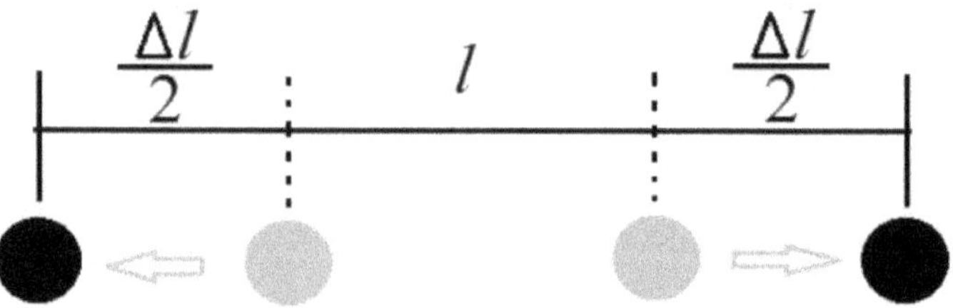

Figure 4-1.   Illustration of strain "h"

The mass pair is shown originally spaced by a distance "l" and locally at rest. By impinging a gravitational wave perpendicularly on the sheet of paper, both particles are shifted respectively to the positions marked by black circles. This shift is denoted by $\Delta l/2$, which means a relative shift between the pair of particles is equal to $\Delta l/l \approx h$, where Δl is the change in the spacing between the particles due to the gravitational wave, "l" is the initial distance between the particles, and "h" is the dimensionless amplitude. In reality, the factor h is more complex and depends on the geometry of the measurement device, the arrival direction, and the frequency and polarisation of the gravitational wave. Nature sets a natural amplitude of h in the order of 10^{-21}.

Comparison of Gravitational Waves with Electromagnetic Waves

There is an enormous difference between gravitational waves and the electromagnetic waves on which our present knowledge of the Universe is based:

Electromagnetic waves are oscillations of the electromagnetic field that propagate through spacetime; gravitational waves are oscillations of the fabric of spacetime itself.

Astronomical electromagnetic waves are almost always incoherent superpositions of emissions from individual electrons, atoms, or molecules.

Cosmic gravitational waves are produced by coherent, bulk motions of huge amounts of mass energy—either material mass or the energy of vibrating, nonlinear spacetime curvature.

Since the wavelengths of electromagnetic waves are small compared to their sources (gas clouds, stellar atmospheres, accretion disks, etc), we can make pictures of the sources. The wavelengths of cosmic gravitational waves are comparable to or larger than their coherent, bulk-moving sources, so we cannot make pictures from them.

The gravitational waves are like sound; they carry, in two independent waveforms, a stereophonic symphony-like description of their sources.

Electromagnetic waves are easily absorbed, scattered, and dispersed by matter. Gravitational waves travel nearly unscathed through all forms and amounts of intervening matter.

Astronomical electromagnetic waves have frequencies that begin at $f \sim 10^7$ Hz and extend upwards by roughly 20 orders of magnitude. Astronomical gravitational waves should begin at $\sim 10^4$ Hz (1000-fold lower than the lowest-frequency astronomical electromagnetic waves) and should extend downwards from there by roughly 20 orders of magnitude.

These enormous differences make it likely that:

The information brought to us by gravitational waves will be very different from that carried by electromagnetic waves; gravitational waves will show us details of the bulk motion of dense concentrations of energy, whereas electromagnetic waves show us the thermodynamic state of optically thin concentrations of matter.

Most (but not all) gravitational-wave sources that our instruments detect will not be seen electromagnetically, and conversely, most objects observed electromagnetically will never be seen gravitationally.

Typical electromagnetic sources are stellar atmospheres, accretion disks, and clouds of interstellar gas, none of which emit significant gravitational waves, while typical gravitational-wave sources are the cores of supernovae (which are hidden from electromagnetic view by dense layers of surrounding stellar gas) and colliding black holes (which emit no electromagnetic waves at all).

Gravitational waves and electromagnetic waves are both predominantly of a particular multipolar form (quadrupolar in the gravitational case and dipolar in the electromagnetic case), but the similarities don't end there. Both types of waves are also transverse to the direction of propagation. That is, the gravitational-wave perturbations of the gravitational field oscillate perpendicular to the direction of propagation, just like electric and magnetic fields. Historically speaking, because of the fundamental limitation of both Maxwell's and Einstein's theories: the fields of the theories cannot travel faster than the speed of light.

Gravitational waves are produced by any mass-energy configuration having a time-varying quadrupole moment and carrying energy and momentum along with them. The term "quadrupole moment" refers to a mathematical concept used in physics, particularly in the study of electromagnetism and General Relativity. It describes the distribution of charge or mass within an object and how that distribution affects the object's response to external forces or fields.

In electromagnetism, the quadrupole moment is related to the distribution of charge within an object and how that distribution generates electric fields. It is a measure of the object's deviation from spherical symmetry. Just as a dipole moment represents a pair of opposite charges separated by a distance, a quadrupole moment involves higher-order terms in the charge distribution.

In the context of General Relativity, the quadrupole moment is related to the distribution of mass in a gravitational system. It plays a crucial role in describing the emission of gravitational waves from asymmetrically accelerating masses, such as binary star systems or spinning compact objects like neutron stars or black holes.

Mathematically, the quadrupole moment tensor is a rank-2 tensor that characterises the distribution of mass or charge within an object. It involves components that describe the distribution in different directions and is related to the second-order derivatives of the mass or charge density. The quadrupole moment tensor describes the shape and orientation of the distribution and plays a crucial role in the generation and emission of gravitational waves. Understanding the quadrupole moment is important in various fields of physics, as it provides insights into the structure and behaviour of physical systems under the influence of gravitational or electromagnetic forces.

Gravitational waves pass through almost everything unaffected, hence carry important information about the origin and nature of gravity. But they distort the spacetime attached to the objects through which they pass. Distances between objects increase and decrease rhythmically at a frequency corresponding to that of the wave as the wave passes. This happens despite the fact that the objects are not acted upon by any unbalanced force. The magnitude of this effect decreases inversely as the distance from the source.

Frequency Bands, Sources, and Detection Methods

Four gravitational-wave frequency bands are being explored experimentally: the high-frequency band (HF; $f \sim 10^4$ to 1 Hz), the low-frequency band (LF; $f \sim 1$ to 10^{-4} Hz), the very-low-frequency band (VLF; $f \sim 10^{-7}$ to 10^{-9} Hz), and the extremely low-frequency band (ELF; $f \sim 10^{-15}$ to 10^{-18} Hz).

1.   High-Frequency Band, 1 to 10^4 Hz

A gravitational wave source of mass M cannot be much smaller than its gravitational radius, $2GM/C^2$, and cannot emit strongly at periods much smaller than the light travel time $4\pi GM/C^3$ around this gravitational radius. Correspondingly, the frequencies at which it emits are

$f = 10^4$ Hz M0/M where M0 is the mass of the Sun, G is the Universal gravitational constant, and c is the speed of light. To achieve a size of the same order as its gravitational radius and thereby emit near this maximum frequency, an object presumably must be heavier than the Chandrasekhar limit. The Chandrasekhar limit is the maximum mass that a white dwarf star can have before it collapses into a neutron star or black hole. The limit is approximately 1.4 times the mass of the Sun or 1.4 M0. Thus, the highest frequency expected for strong gravitational waves is fmax $\sim 10^4$ Hz. This defines the upper edge of the high-frequency gravitational wave band.

The high-frequency band is the domain of the Earth-based gravitational-wave detectors of laser interferometers and resonant mass antennae. At frequencies below about 1 Hz, Earth-based detectors face nearly insurmountable noise due to the atmosphere and other vibrations. This defines the 1 Hz lower edge of the high-frequency band; to detect waves below this frequency, one must fly one's detectors in space.

2.   Low-Frequency Band, 10^{-4} to 1 Hz

The low-frequency band, 10^{-4} to 1 Hz, is the domain of detectors operating in space (in Earth orbit or in interplanetary orbit). The 1 Hz upper edge of the low-frequency band is defined by the gravity gradient and seismic cutoffs on Earth-based instruments; the 10^{-4} Hz lower edge is defined by expected severe difficulties at lower frequencies in isolating spacecraft from the buffeting forces of fluctuating solar radiation pressure, solar wind, and cosmic rays. The low-frequency band should be populated by waves from short-period binary stars in our own galaxy: from white dwarfs, neutron stars, and small black holes spiralling into massive black holes.

3.   Very-Low-Frequency Band, 10^{-7} to 10^{-9} Hz

The only compact bodies that can radiate in the VLF band or below, i.e., at $f \leq 10^{-7}$ Hz, are those with $M \geq 10^{11}$ M0 and generally are not much larger than their own gravitational radii.

Millisecond pulsars are used to study the gravitational waves in the VLF band. By averaging the pulses' times of arrival over long periods of time (months to tens of years), a very high timing precision can be achieved, and correspondingly, tight limits can be placed on the waves bathing the Earth or the pulsar. Conventional astronomical wisdom suggests that compact bodies this massive (10^{11} M0) do not exist and therefore, the only strong waves in the VLF band and below are a stochastic background produced by the same early-Universe processes that might radiate at low and high frequencies: cosmic strings, phase transitions, and the Big Bang.

4.   Extremely-Low-Frequency Band, 10^{-15} to 10^{-18} Hz

Gravitational waves in the extremely low-frequency band (ELF), 10^{-15} to 10^{-18} Hz, should produce anisotropies in the cosmic microwave background radiation. This leads to an intriguing situation whereby the gravitational wavelength is about π times the Hubble distance and the waves, by squeezing all of the space inside our cosmological horizon in one direction and stretching it all in another, should produce a quadrupolar anisotropy in the microwave background.

The quadrupolar anisotropy is measured by the Cosmic Background Explorer (COBE) satellite. NASA's COBE (Cosmic Background Explorer) satellite rocketed into Earth orbit on Nov. 18, 1989, and quickly revolutionised our understanding of the early cosmos. COBE data refined our knowledge of the CMB (cosmic microwave background), the oldest light in the Universe, and provided key evidence supporting the Big Bang Theory as an explanation for the origin of the Universe. COBE investigators John Mather and George Smoot were awarded the 2006 Nobel Prize in physics for this work. COBE was retired on Dec. 23, 1993.

Gravitational Wave Polarizations

Gravitational waves exist in two polarisation states, namely plus (+) and cross (×) polarisations. The angles between these polarisations are $\pi/4$ rather than $\pi/2$ as in the case of electromagnetic waves. Different polarisations have different effects on an array of particles. A circular ring of free test particles in the x-y plane suffers tidal deformation into an elliptical ring of the same area with the major axis in the x and y directions alternatively by a plus (+) polarised wave, whereas a cross (×) polarised gravitational wave deforms the ring into an ellipse tilted at an angle of $\pi/4$ (fig. 1).

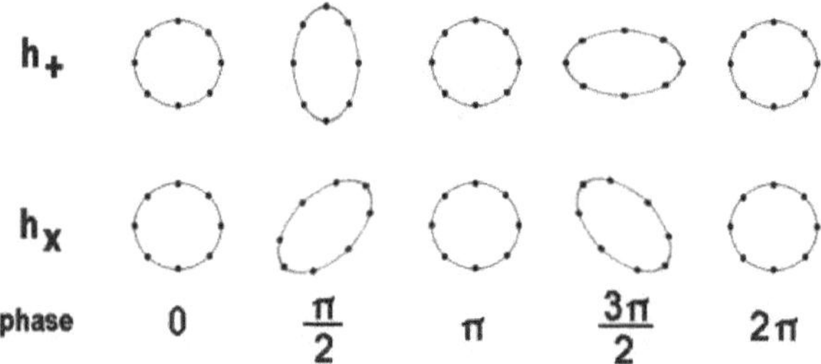

Figure 4-2. Plus (+) polarised waves and Cross (×) polarised gravitational waves

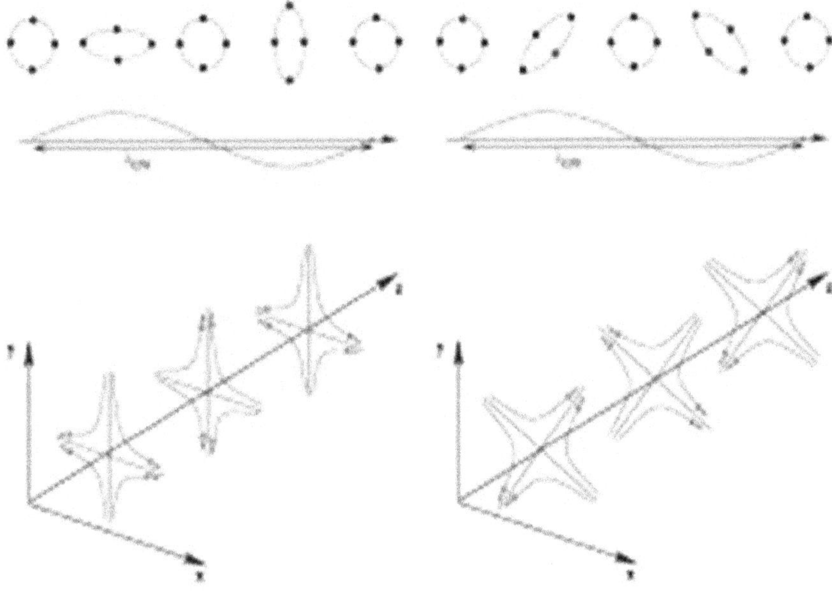

Figure 4-3. Depicting the effects of polarisation of gravitational waves on the test particles in a ring.

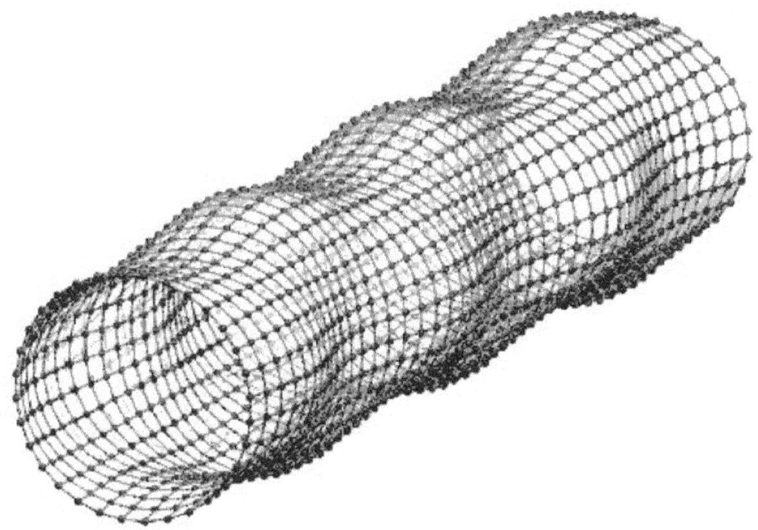

Figure 4-4. Gravitational waves propagating and warping the space-time continuum.

Gravitational waves propagate in one direction, alternately expanding and compressing space in mutually perpendicular directions, defined by the gravitational wave's polarization.

These polarisation states are crucial because they determine how gravitational waves interact with detectors like LIGO (Laser Interferometer Gravitational-Wave Observatory) or Virgo. By measuring the effects of gravitational waves in different directions and polarisation states, scientists can infer various properties of the sources that produce these waves, such as the masses and orientations of merging black holes or neutron stars.Top of Form

After consideration of the above-mentioned properties, one can summarise the characteristics of gravitational waves as follows:

1. Frequency and Amplitude: Gravitational waves have a frequency and amplitude. The frequency determines the number of oscillations per unit of time, while the amplitude represents the strength or intensity of the wave.

2. Transverse Nature: Gravitational waves are transverse waves, meaning that the oscillations occur perpendicular to the direction of wave propagation. This is similar to electromagnetic waves.

3. Traceless Quadrupole Moment: Gravitational waves are predominantly quadrupolar in nature. This means that they are generated by the time variation of the quadrupole moment of the source. The quadrupole moment is a mathematical representation of the distribution of mass or energy in the source.

4. Stretching and Compressing Spacetime: Gravitational waves cause the stretching and compressing of spacetime as they

pass through. This is due to the fact that they carry energy and momentum, which interact with the curvature of spacetime.

5. Weak Interaction with Matter: Gravitational waves interact very weakly with matter, making them difficult to detect. This is because they have a very small amplitude and do not easily transfer energy to particles or objects.

6. Speed of Light: Gravitational waves travel at the speed of light in a vacuum. This is a fundamental property of waves in General Relativity.

7. Open New Observational Windows: Gravitational waves provide a new way to observe and study the Universe. They can reveal information about the dynamics of astrophysical events and phenomena that are not accessible through other means, such as electromagnetic radiation.

The GW Spectrum

Whereas astrophysical electromagnetic waves are incoherent and have wavelengths typically much smaller than their sources, ranging from a few kilometres down to sub-nuclear wavelengths. Gravitational waves are coherent and have wavelengths larger than their sources, with wavelengths starting at a few kilometres and ranging up to the size of the Universe.

The Gravitational-Wave Spectrum

Much like electromagnetic waves, gravitational waves are emitted by many different objects over a wide range of frequencies. Terrestrial interferometers such as the Laser Interferometer Gravitational-wave Observatory (LIGO) and Virgo are sensitive to only a subset of those frequencies, which limits their ability to "see" certain cosmic phenomena. They will not detect collisions of supermassive black holes found in the hearts of galaxies, for example. But space-based interferometers and other approaches for picking up gravitational waves could extend physicists' reach.

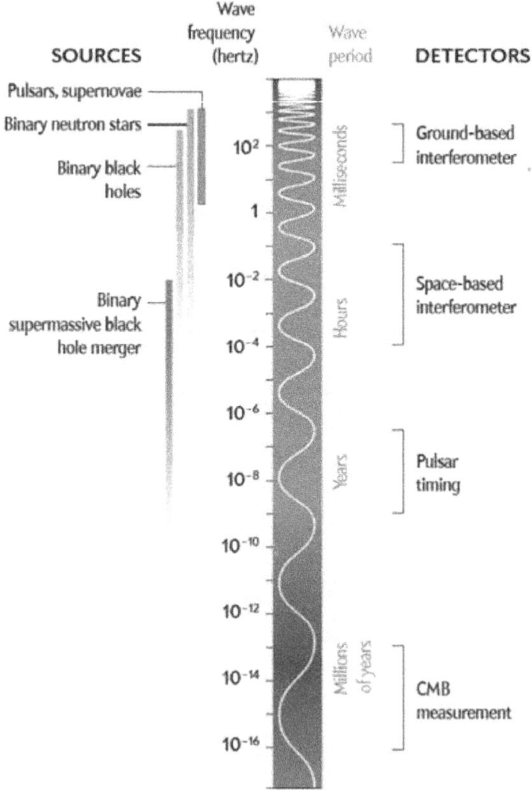

CHAPTER – 5

Detection of Gravitational Waves By Ligo

LIGO opens a new window on the Universe with the observation of gravitational waves from colliding black holes. LIGO stands for "Laser Interferometer Gravitational-Wave Observatory". It is the world's largest gravitational wave observatory and a marvel of precision engineering.

For the first time, scientists have observed ripples in the fabric of spacetime called gravitational waves, arriving at Earth from a cataclysmic event in the distant Universe. This confirms a major prediction of Albert Einstein's 1915 General Theory of Relativity and opens an unprecedented new window to the cosmos. Physicists have concluded that the detected gravitational waves were produced during the final fraction of a second of the merger of two black holes to produce a single, more massive spinning black hole. This collision of two black holes had been predicted but never observed.

Albert Einstein was convinced it would never be possible to measure gravitational waves and was unsure whether the waves were real or just a mathematical illusion. His contemporary colleague, Arthur Eddington, was even more sceptical and pointed out that gravitational waves appeared "to propagate at the speed of thought".

Setting aside that scepticism, gravitational waves were detected on 14th September 2015 by both of the twin Laser Interferometer Gravitational-Wave Observatory (LIGO) detectors, located in Livingston, Louisiana, and Hanford, Washington. The LIGO observatories are funded by the National Science Foundation (NSF) and were conceived, built, and operated by the California Institute of Technology (Caltech) and the Massachusetts Institute of Technology (MIT).

Based on the observed signals, LIGO scientists estimate that the black holes for this event were about 29 and 36 times the mass of the Sun, and the event took place 1.3 billion years ago. About three times the mass of the Sun was converted into gravitational waves in a fraction of a second with a peak power output about 50 times that of the whole visible Universe. By looking at the time of arrival of the signals, the detector in Livingston recorded the event 7 milliseconds before the detector in Hanford, scientists can say that the source was located in the Southern Hemisphere.

According to General Relativity, a pair of black holes orbiting around each other lose energy through the emission of gravitational waves, causing them to gradually approach each other over billions of years, and then much more quickly in the final minutes. During the final fraction of a second, the two black holes collide at nearly half the speed of light and form a single, more massive black hole, converting a portion of the combined black holes' mass to energy, according to Einstein's formula $E = mc^2$. This energy is emitted as a final strong burst of gravitational waves.

LIGO

LIGO is no ordinary telescope for detecting light and other electromagnetic radiation from space. It is an instrument for listening to space's gravitational waves; even though gravitational waves are tremors in spacetime itself and not sound waves, their frequency is equivalent to those we can hear with our human ears.

Though it's called an observatory, LIGO is unlike any other observatory on Earth. Ask someone to draw a picture of an observatory and odds are it will look something like the photo below: a typical telescope dome on a mountain top.

As a gravitational wave observatory, LIGO bears no resemblance to this whatsoever, as the photo of the LIGO Livingston interferometer below clearly illustrates.

Figure 5-1. Aerial photo of LIGO Livingston, Louisiana, showing all of one 4-km-long arm and part of the other (off to the right). The visible arms are concrete structures that protect the vacuum tubes from the elements. (Credit: Caltech/MIT/LIGO Lab)

Although LIGO will search for gravitational waves from space, and it is called an "observatory", LIGO is not, strictly speaking, an astronomical facility. LIGO is truly a physics experiment on the scale and complexity of some of the world's giant particle accelerators and nuclear physics laboratories. Though its mission is to detect gravitational waves from some of the most violent and energetic processes in the Universe, the data it will collect will have far-reaching effects on many areas of physics including gravitation, relativity, astrophysics, cosmology, particle physics, and nuclear physics.

Since LIGO has the word "observatory" in it, however, it is helpful to first describe how it differs from the observatories that most people envision. Three physical differences significantly distinguish LIGO from an astronomical observatory:

First, LIGO is blind. Unlike optical or radio telescopes, LIGO does not see electromagnetic radiation (e.g. visible light, radio waves, microwaves). It doesn't have to because gravitational waves are not part of the electromagnetic spectrum. They are a completely different phenomenon altogether. In fact, electromagnetic radiation is so unimportant to LIGO that its detector components are completely isolated and sheltered from the outside world.

Second, LIGO is the opposite of round. Since LIGO doesn't need to focus light or radio waves from stars or other objects in the Universe, it doesn't need to be dish-shaped like telescope mirrors or radio dishes, which focus electromagnetic radiation into images. Instead, LIGO's eyes on the sky are more like ears. LIGO's ears consist of two perfectly straight and level 4 km (2.5-mile) long steel vacuum tubes, 1.2m in diameter, arranged in the shape of an "L", and protected by a 10-foot-wide, 12-foot-tall concrete enclosure that protects the vacuum tubes from the outside world.

Third, LIGO cannot function alone. While an astronomical observatory can function and collect data just fine on its own (some don't by choice), gravitational wave observatories like LIGO cannot operate solo. The only way to definitively detect a gravitational wave is by operating in unison with a distant twin so that local vibrations are not mistaken for signals from gravitational waves.

LIGO COMES INTO BEING

LIGO was originally proposed in the 1980s by three United States emeritus professors of physics, Rainer Weiss of MIT and RP Feynman and Ronald Drever of Caltech. The work started on LIGO after initial funding was approved by the National Science Foundation (NSF), USA, in 1992.

LIGO observatories are established in the USA at Livingston, Louisiana, and at Hanford, Washington, 3030 km apart, working in unison. This distance corresponds to a difference in gravitational waves' arrival times of up to 10 ms.

Figure 5-2. Aerial views showing the locations and extents of the LIGO Hanford and LIGO Livingston interferometers.

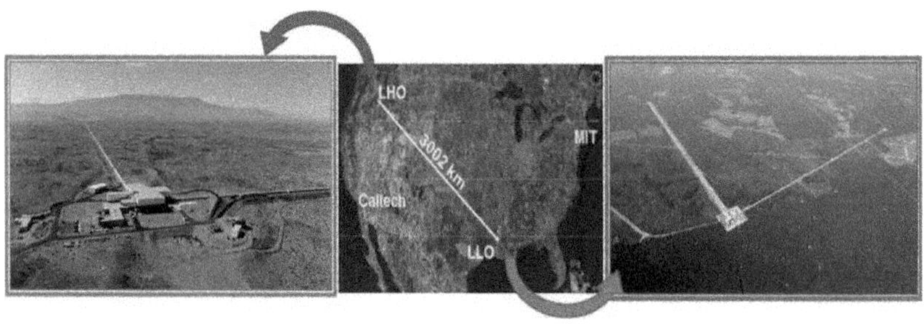

Figure 5-3.   The general locations of the LIGO Hanford and LIGO Livingston interferometers. (Caltech/MIT/LIGO Lab)

Both sites are relatively seismically quiet, with low human noise.

LIGO HANFORD OBSERVATORY (LHO), Hanford,

- located on the DOE reservation
- treeless, semi-arid high desert
- 25 km from Richland, WA

LIGO LIVINGSTON OBSERVATORY (LLO), Livingston, LA

- located in a forested, rural area
- Commercial logging in a wet climate
- 50 km from Baton Rouge, LA

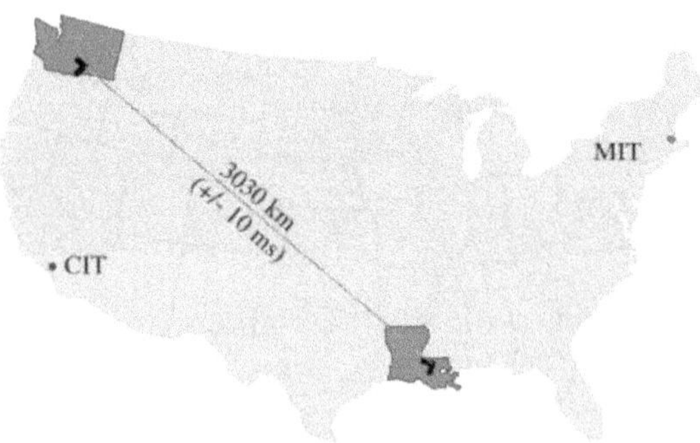

Figure 5-4.   Location of two observatories of LIGO

MIT: Massachusetts Institute of Technology

CIT: California Institute of Technology

These instruments act as "antennae" to detect gravitational waves. The LIGO interferometer is basically a Michelson interferometer, 360 times greater than the original interferometer of the Michelson-Morley experiment. This increased length enhances the sensitivity of the instrument. The initial LIGO operations took place between 2002 and 2010 and could not detect any gravitational waves, showing that the sensitivity of the instrument was still inadequate to observe a change in length of the order of 10^{-20} as required for the gravitational wave's detection.

What is an Interferometer?

Interferometers are investigative tools used in many fields of science and engineering. They are called interferometers because they work by merging two or more sources of light to create an interference pattern, which can be measured and analysed; hence 'Interfere-o-meter', or interferometer. The interference patterns generated by interferometers contain information about the object or phenomenon being studied. They are often used to make very small measurements that are not achievable in any other way. This is why they are so powerful for detecting gravitational waves—LIGO's interferometers are designed to measure a distance 1/10,000th the width of a proton!

Interferometers were pioneered in the mid-to-late 1800s as many scientists made attempts to disprove the existence of luminiferous ether, proposed most famously by Augustin-Jean Fresnel. Among this international group of scientists were American physicists Albert Michelson and Edward Morley, who invented a self-named optical configuration, the Michelson-Morley Interferometer. Results from their experiments published in 1887 are often cited as the first conclusive experimental evidence against the existence of the ether.

As a part of common knowledge and wide use almost a century later in the late 1960s, their particular interferometric configuration was identified as a natural fit for the detection of gravitational wave strain on spacetime, given its precise measurement of the phase change within two perpendicular arms. As such, the Michelson interferometer's optical configuration is a core, critical piece of all of today's gravitational wave detectors, including LIGO.

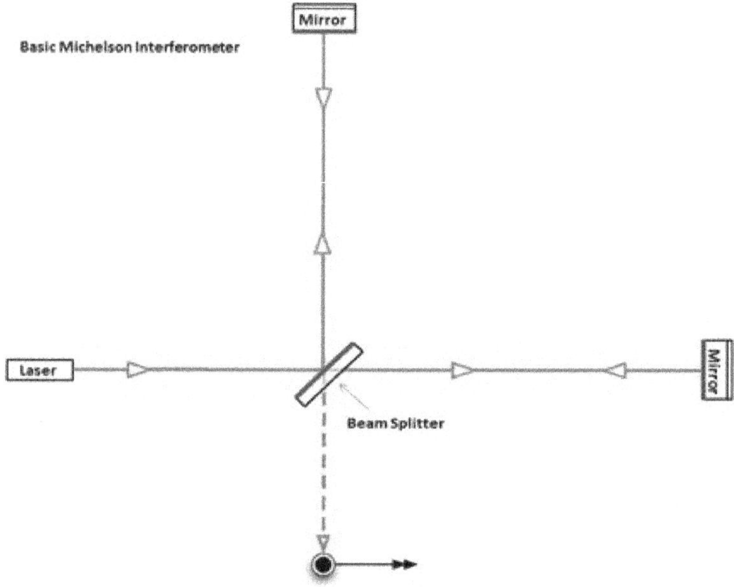

Figure 5-5. *The basic configuration of a Michelson Laser Interferometer is shown above. It consists of a laser, a beam splitter, a series of mirrors, and a photodetector (the black dot) that records the interference pattern.*

Despite their different designs and the various ways in which they are used, all interferometers have one thing in common: they superimpose beams of light to generate an interference pattern.

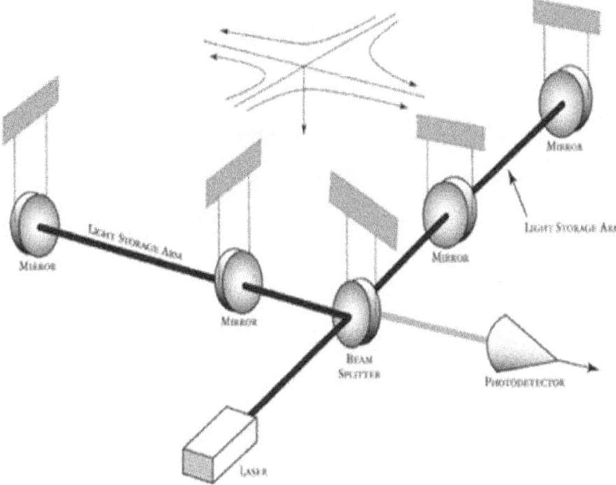

Figure 5-6.   Basic schematic of LIGO's interferometers with an incoming gravitational wave depicted as arriving from directly above the detector. (Caltech/MIT/LIGO Lab)

Detection of Gravitational Waves by LIGO's Interferometer

Gravitational waves cause space itself to stretch in one direction and simultaneously compress in a perpendicular direction. In LIGO, this causes one arm of the interferometer to get longer while the other gets shorter, then vice versa, back and forth as long as the wave is passing. The technical term for this motion is "Differential Arm" motion, or differential displacement since the arms are simultaneously changing lengths in opposing ways, or differentially.

As described above, as the lengths of the arms change, so too does the distance travelled by each laser beam. A beam in a shorter arm will return to the beam splitter before the beam in a longer arm, then the situation switches as the arms oscillate between being longer and shorter. Arriving at different times, the waves of light no longer meet up nicely when recombined at the beam splitter. Instead, they shift in and out of alignment or "phase" as they merge while the wave is causing the arm lengths to oscillate, manifested as a "Flicker" at the detector.

While in principle the idea seems almost simple, in practice, detecting that flicker is not. The change in arm length caused by a gravitational wave can be as small as 1/10,000th the width of a proton (that's 10^{-19} m)! Furthermore, finding a gravitational wave flicker among all the other flickers LIGO experiences (caused by anything that can shake the mirrors, like earthquakes or traffic on nearby roads) is another story.

The first and most obvious difference between a typical Michelson interferometer and LIGO's interferometers is its scale. With arms 4 km (2.5 miles) long, LIGO's interferometers are by far the largest ever built. (By contrast, the interferometer Michelson and Morley used in their famous experiment to study the "aether" had arms about 1.3m long). The scale of LIGO's instruments is crucial to its search for gravitational waves. The longer the arms of an interferometer, the smaller the measurements they can make. And having to measure a change in distance 10,000 times smaller than a proton means that LIGO has to be larger and more sensitive than any interferometer ever before constructed. While 4-km-long arms seem pretty huge, if LIGO's interferometers were simple Michelsons, they would still be too short to enable the detection of gravitational waves. And of course, there are practical limitations to building such a precision instrument much larger than 4 km (see Chapter 7: Roadmap). So how can LIGO possibly make the measurements it does?

Longer is Better

This paradox of limiting the length of the interferometer arm to 4 km was solved by altering the design of the Michelson to include something called "Fabry-Perot cavities". The figure below shows a basic Michelson design modified to include such cavities. An additional mirror is placed in each arm near the beam splitter (the box on the 45-degree angle), 4 km from the mirror at the end of that arm. This 4-km-long space comprises the Fabry-Perot cavity. After entering the instrument via the beam splitter, the laser in each arm bounces between its two mirrors about 300 times before being merged with the beam from the other arm. These reflections serve two functions:

1. It builds up the laser light within the interferometer, which increases LIGO's sensitivity (more photons also make LIGO more sensitive).

2. It increases the distance travelled by each laser from 4 km to 1200 km, thereby solving the length problem.

Since we know that the longer the arms of an interferometer, the more sensitive the instrument is to vibration, this design significantly increases LIGO's sensitivity and enables it to detect changes in arm length much smaller than a proton—the size of changes expected to be caused by a gravitational wave. And thanks to Fabry-Perot cavities, LIGO can achieve this sensitivity with arms just 4 km long.

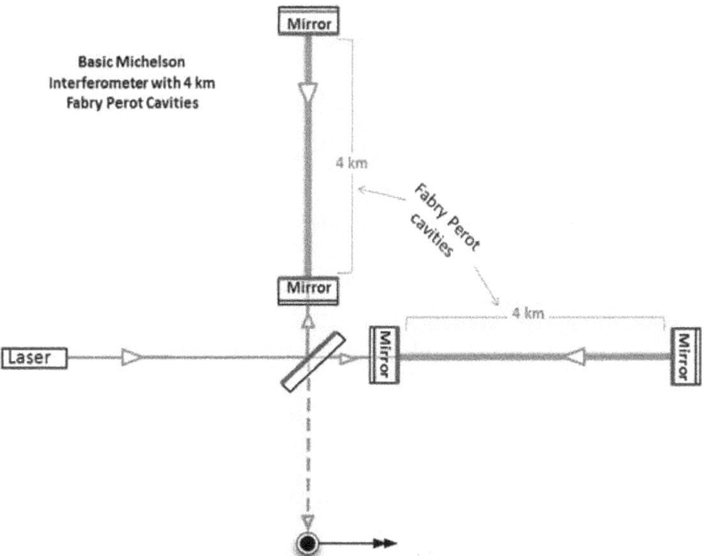

Figure 5-7. Basic Michelson interferometer with Fabry-Perot cavities. Additional mirrors are inserted near the beam splitter to facilitate multiple reflections of the laser, containing it within the interferometer and increasing the distance travelled by the beams. This greatly increases LIGO's sensitivity to the smallest changes in arm length. (Caltech/MIT/LIGO Lab)

LIGO Needs More Power!

Length isn't the only limiting factor in LIGO's sensitivity. Laser power is also a consideration. Just as increasing length increases the interferometer's sensitivity, increasing laser power also enhances its performance. While increasing length amplifies tiny changes in arm length, increasing laser power results in increasing the interferometer's resolving power. In other words, the more laser photons emerge from each arm, the sharper the fringes that are measured by the photodetector.

But there's a problem here too. LIGO's laser beam first enters the interferometer at about 40 Watts, but it needs to operate closer to 750 kW if it has any hope of detecting gravitational waves. Here, there is another paradox. Just as it would be impossible to build a 1200-km-long interferometer, building a laser with this initial power is a practical impossibility.

Once again, LIGO uses mirrors to solve this dilemma. They are called Power Recycling Mirrors. The figure below shows a power recycling mirror located inside each interferometer.

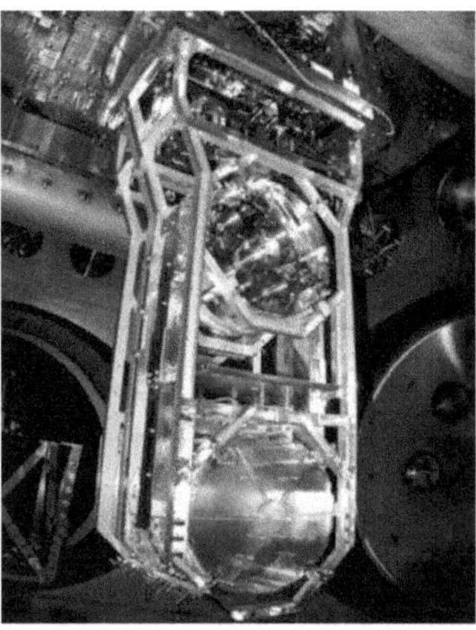

Figure 5-8. Power Recycling Mirror of LIGO (Courtesy of Caltech/MIT/ LIGO Laboratory)

Each mirror in LIGO's optical cavities sits at the end of a so-called quadruple pendulum. The 40 kg mirror is not attached to the outside frame but hangs from the 40 kg silica cylinder above it by a set of fused silica fibres. The cylinder in turn hangs from a metal mass, which is hanging from yet another suspended metal mass. The system is designed to minimise thermal noise and passively dampen any environmental vibrations that get past LIGO's active seismic isolation system.

Inside the interferometer, light from the laser passes through the transparent side of a power recycling mirror to the beam splitter and then onto the arms of the interferometer. The beam splitter is a partially polished mirror through which half the light comes out; the other half is reflected back into the room. The instrument's alignment, mirror coatings, and even quantum mechanics ensure that nearly all of the laser light entering the arms follows a path back to the reflective side of the power recycling mirror rather than to the photodetector. As laser power is constantly entering the interferometer, the power recycling mirror continually reflects the laser light that has travelled through the instrument back into the interferometer (hence recycling). This process greatly boosts the power of the laser beam inside the Fabry-Perot cavities without the need to generate such a powerful laser beam at the outset.

The boost in power generated by power recycling results in a sharpening of the interference fringes that appear when the two beams are superimposed—fringes that will tell scientists whether a gravitational wave has passed or not. The sharper the fringes, the easier it becomes to identify the tell-tale signs of gravitational waves.

LIGO's interferometers were constructed with extraordinary mechanisms to dampen out unwanted vibrations (noise), making it easier for scientists to weed out vibrations caused by gravitational waves.

With these modifications, LIGO's interferometer is known as a Dual Recycled Fabry-Perot Michelson; but at its heart, it is still a Michelson interferometer.

In the end, though, the name of the game is a signal to noise. At the high-frequency range of LIGO's detection band (10 Hz to 7000 Hz), the dominant source of noise is quantum fluctuations in the detected photon arrival rate. Because LIGO is configured for near-complete destructive interference at the photodetector, almost all the light ends up reflected back towards the laser source. LIGO incorporates mirrors that recycle that light back into the system. At lower frequencies (below 150 Hz), seismic, thermal, and radiation-pressure noise come into play. Eliminating as much of that noise as possible was crucial to LIGO's success.

The observatory became operational in 2002 and started searching for evidence of gravitational waves. After five searches, the first phase of LIGO ended in 2005 with no detection of gravitational waves. The sensors were improved, and the Enhanced LIGO commenced the hunt for gravitational waves in 2009, which again conducted the search in vain until 2010.

Advanced LIGO (aLIGO)

Between 2010 and 2014, the instruments underwent a complete overhaul to become Advanced LIGO. Improvements included the installation of an active seismic isolation system and the replacement of the 11 kg mirrors in the optical cavities with 40 kg mirrors for reduced thermal noise and radiation-pressure noise. In addition, the mirrors were hung from a quadruple pendulum system to better dampen any mechanical vibrations that might get past the active system.

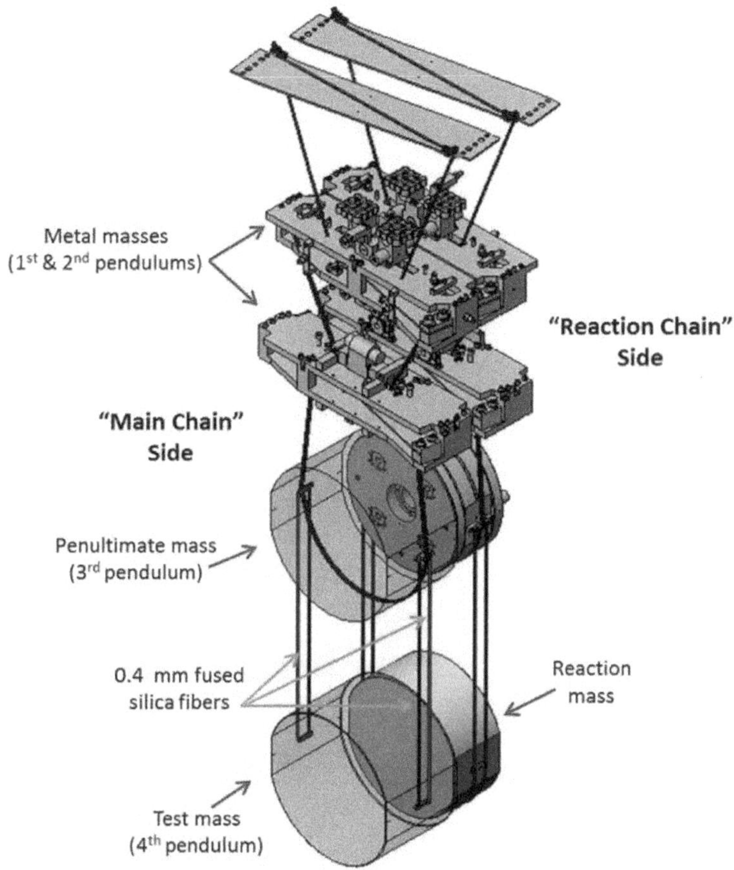

Figure 5-9.   Anatomy of a quad. Four vibration-damping masses are present. The top two are flat metal structures below which the cylindrical masses hang from 0.4mm silica fibres. The bottom two on the Main Chain side are cylindrical and made of high-quality fused silica glass. (Adapted from IGR, University of Glasgow)

Passive Damping

LIGO's passive damping system holds the all-important mirrors perfectly still through a 4-stage pendulum called a "quad". In the quad, LIGO's test masses (its mirrors) are suspended at the end of four pendulums by 0.4 mm thick fused silica (glass) fibres. The "Main Chain" side faces the laser beam, while the "Reaction mass" side helps to keep the test mass steady from noise not associated with sources from space. This configuration absorbs any movement not completely concealed by the active system. The sheer weight of the suspension components (each mirror weighs 40 kg) also helps to prevent motion of the mirrors thanks to the Law of Inertia.

Working together, these active and passive vibration-damping systems ensure that LIGO's lasers and mirrors are isolated from as much external noise and vibration as is physically possible.

When LIGO detected GW150914, it was running in low-power mode. By 2019, when the upgraded instruments are fully commissioned and operating at full power, Advanced LIGO should have a sensitivity 10 times that of the initial LIGO detectors. The tenfold increase in sensitivity meant the new and Advanced LIGO could 'listen' for gravitational waves 10 times farther away than the initial LIGO. In other words, the Advanced LIGO would have access to a thousand times more volume of space.

Experimental Results

LIGO can convert its spacetime distortion signals into an audible sound called a "chirp" so we can all, in a sense, 'hear' the final moments of the lives of two black holes, two neutron stars, or a black hole-neutron star merger. These objects have been orbiting each other for billions of years; LIGO captures the last fraction of a second or a few seconds of that lifetime together.

This cataclysmic event, producing the gravitational-wave signal GW150914, took place in a distant galaxy more than one billion light-years from the Earth.

This GW discovery provides the first robust confirmation of several theoretical predictions: (i) that "heavy" Black Holes (BHs) exist, (ii) that binary black holes (BBHs) form in nature, and (iii) that BBHs merge within the age of the Universe at a detectable rate.

LIGO estimated that the peak gravitational-wave power radiated during the final moments of the black hole merger was more than 10 times greater than the combined light power from all the stars and galaxies in the observable Universe. So, the LIGO detectors have observed a truly remarkable event that happened a long time ago in a galaxy far, far away. This remarkable discovery marks the beginning of an exciting new era of astronomy as we open an entirely new, gravitational-wave window on the Universe.

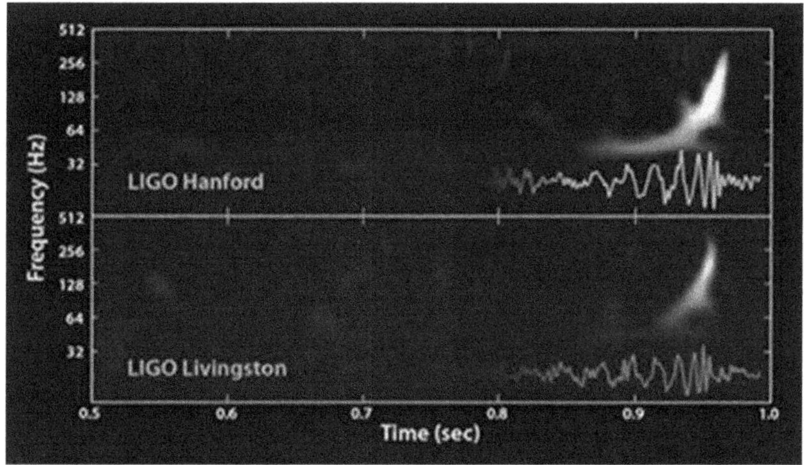

Figure 5-10. The gravitational wave event GW150914 observed by the LIGO Hanford Observatory-LHO (top panel) and LIGO Livingston Observatory-LLO (bottom panel) detectors. The two plots show how the gravitational wave strain produced by the event in each LIGO detector varied as a function of time (in seconds) and frequency (in hertz, or the number of wave cycles per second). Both plots show the frequency of GW150914 sweeping sharply upwards, from 35 Hz to about 150 Hz over two-tenths of a second. GW150914 arrived first at LLO and then at LHO about seven-thousandths of a second later—consistent with the time taken for light, or gravitational waves, to travel between the two detectors.

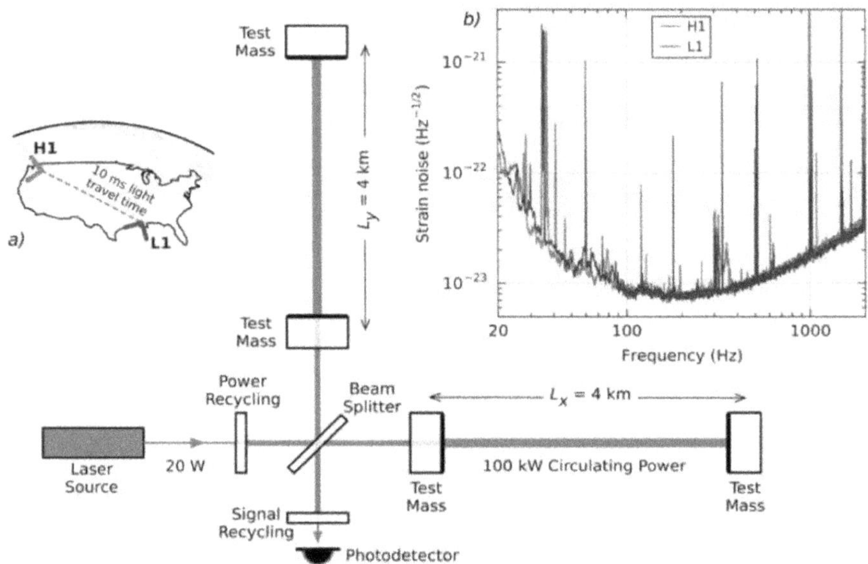

*Figure 5-11. Simplified diagram of an Advanced LIGO detector (not to scale), including several of the key enhancements to the basic design: an **optical cavity** that reflects the laser light back and forth many times in each arm, multiplying the effect of the gravitational wave on the phase of the laser light; a **power recycling** mirror that increases the power of the laser in the interferometer as a whole; a **signal recycling** mirror that further optimises the signal extracted at the **photodetector**. These enhancements boost the power of the laser in the optical cavity by a factor of 5000 and increase the total amount of time that the signal spends circulating in the interferometer.*

Inset (a), on the left, shows the locations and orientations of the two LIGO observatories and indicates the light travel time between them. Inset (b) shows how the instrument strain noise varied with frequency in each detector near the time of the event. The lower the instrument noise, the higher the detectors' sensitivity. The tall spikes indicate narrow frequency ranges where the instrument noise is particularly large.

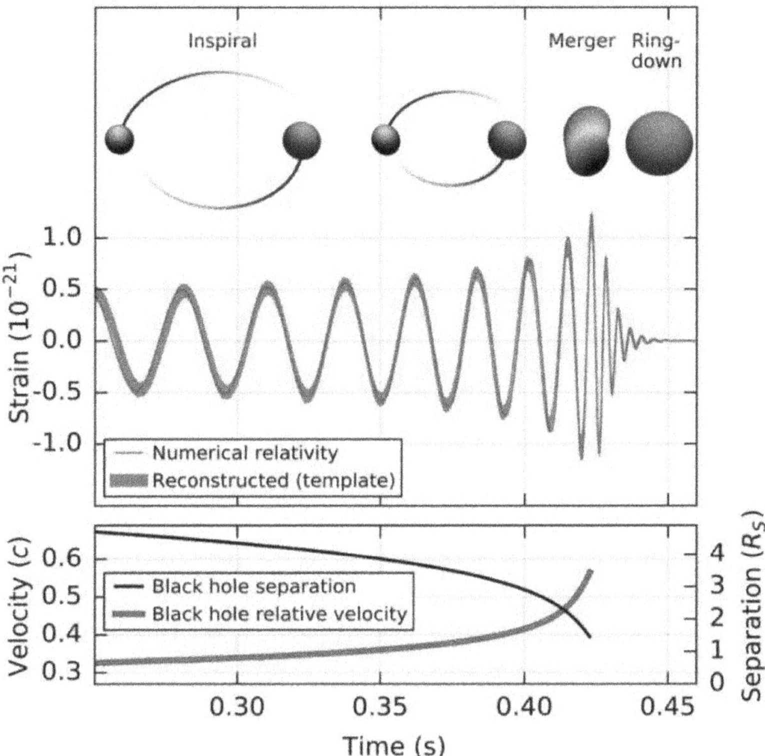

Figure 5-12. Some key results of the analysis of GW150914, comparing the reconstructed gravitational-wave strain (as seen by LHO at Hanford) with the predictions of the best-matching waveform computed from General Relativity over the three stages of the event: inspiral, merger, and ringdown. Also shown are the separation and velocity of the black holes and how they change as the merger event unfolds.

The statistical significance of the signal is very high, exceeding the "five-sigma" standard that physicists use to distinguish evidence strong enough to claim a discovery.

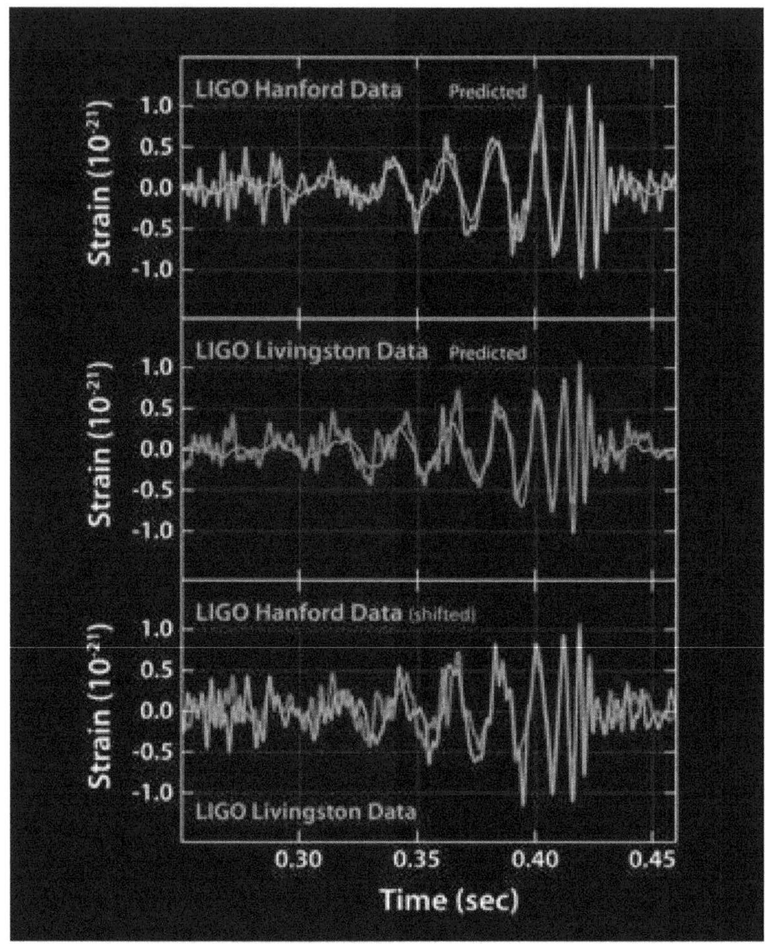

Figure 5-13. GW150914 signal observed by the twin LIGO observatories at Livingston, Louisiana, and Hanford, Washington. The signals came from two merging black holes, each about 30 times the mass of our Sun. The top two plots show data received at Livingston and Hanford, along with the predicted shapes for the waveform. These predicted waveforms show what two merging black holes should look like according to the equations of Albert Einstein's General Theory of Relativity, along with the instrument's ever-present noise. Time is plotted on the X-axis and strain on the Y-axis.

The binary is approximately 1.3 billion light-years from Earth, or equivalently, at a luminosity distance of 400 megaparsecs. A megaparsec (Mpc) is a unit of measurement used to measure distances in space, particularly between galaxies. It is equal to one million parsecs, or 3.26 million light-years. The researchers estimate that about 4.6% of the binary's energy was radiated in gravitational waves, leading to a rotating black hole remnant with a mass 62 times the mass of the Sun and a dimensionless spin of 0.67.

However, all of a sudden, just as the power emitted by these inspiraling black holes reaches a peak, these gravitational waves cease to be emitted altogether. The two black holes have now passed within one another's event horizon, and those ripples in spacetime face the same problem as light: they cannot escape from within the black hole's event horizon. Once both masses, like two black holes (or a neutron star-black hole pair, or a neutron star-neutron star merger that creates a black hole), fall behind the event horizon, that's the end of the game: no more gravitational waves generated from inside the event horizon can get out. All that escapes is gravitational radiation from what's known as the ringdown phase: where the event horizon changes back into a shape that represents a single, final, post-merger object, as opposed to the two distinct objects from before the merger.

The length of time that a signal remains in LIGO's interferometers, and hence the quality of a potential detection of gravitational waves, depends inversely on the frequency that LIGO is set up to measure and the masses of the binary objects involved. It is, therefore, easier to detect gravitational waves at lower frequencies and from lighter objects.

Before its upgrade, LIGO was able to detect gravitational waves from 40 to 10,000 Hz, but since aLIGO came online, the interferometers have been able to detect waves down to a frequency of just 10 Hz, thereby greatly extending LIGO's reach.

Figure 5-14. The 2017 Nobel Prize in Physics has been awarded to three US scientists, Rainer Weiss (Massachusetts Institute of Technology), Kip Thorne (California Institute of Technology), and Barry Barish (California Institute of Technology) for the detection of gravitational waves. These three scientists have been selected because of their decisive contributions to the development of the LIGO detector and observation of gravitational waves.

Second LIGO detection: GW151226 with two masses of black holes of masses $14M_0$ and $8M_0$ and a final black hole mass of $21M_0$, releasing gravitational wave energy of one solar mass. This second detection shows that the initial discovery was not a rare windfall, but rather a preview of many things to come. The main takeaway of this detection is that: (a) this merger created gravitational waves of a higher frequency, (b) gravitational waves were "visible" for a longer period, (c) scientists were able to track the final 27 orbits of the binary before they merged.

LIGO's 3rd observation is GW170114 with binary black hole masses $32M_0$ and $19\ M_0$ with remnant mass $49\ M_0$ (energy released is $2\ M_0$).

This was followed by the first and second detection of gravitational waves from a so-called "mixed merger" consisting of a black hole colliding with and swallowing a neutron star. The events designated

GW200105 and GW200115 were picked up just 10 days apart on 5 Jan 2020 and 15 Jan 2020, respectively.

As of November 2021, the LIGO-Virgo-KAGRA collaboration has detected more than 90 gravitational wave signals from a variety of different events. This included the first detection by LIGO and VIRGO of a "neutron star-neutron star merger" on August 17, 2017, designated GW170817. (Appendix – II)

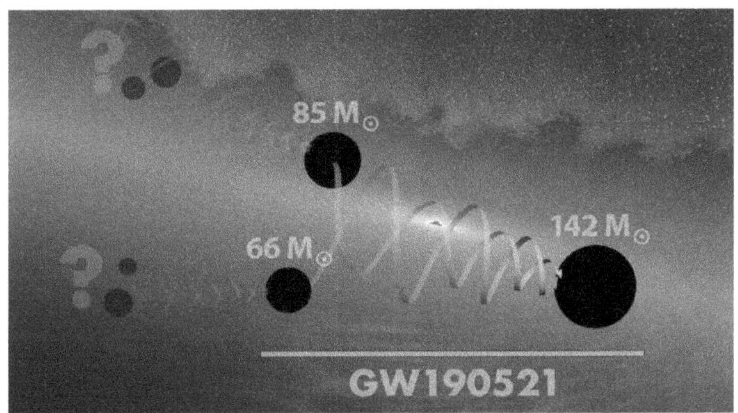

Figure 5-15. The merging black holes that formed the GW190521 event. This was also the first time that intermediate-mass merging black holes had been detected. Credit: LIGO/Caltech/MIT/R Hurt (IPAC)

Scientists deciphering signals from GW190521 were able to identify the two progenitors as black holes but found that they had unexpectedly large masses. Perhaps they were the products of other, smaller black hole collisions aeons ago, or maybe they are the results of some as-yet-unknown physical processes.

By studying the evolution of stars, scientists can better understand how they end up as compact remnants, and this is an important part of predicting what the gravitational waves will look like when the objects collide. This type of research guides the design and construction of the next generation of gravitational wave detectors.

CHAPTER – 6
Virgo and Kagra

VIRGO

The Virgo Interferometer is a large instrument designed to detect gravitational waves predicted by Einstein's General Theory of Relativity. The interferometer is named after the Virgo Cluster of about 1,500 galaxies in the Virgo constellation, about 50 million light-years from Earth.

Figure 6-1.   An aerial view of the Virgo gravitational wave detector. The Virgo detector is located in the countryside of the Commune of Cascina, a few kilometres south of the city of Pisa, Tuscany, Italy.

The VIRGO Collaboration is currently composed of more than 500 scientists, engineers, and technicians from 99 institutes from Belgium,

France, Germany, Hungary, Italy, the Netherlands, Poland, and Spain. The European Gravitational Observatory (EGO) hosts the Virgo detector near Pisa in Italy and is funded by the Centre National de la Recherche Scientifique (CNRS) in France, the Istituto Nazionale di Fisica Nucleare (INFN) in Italy, and Nikhef in the Netherlands.

Its construction was completed in June 2006. Virgo is a recycled Michelson interferometer with 3-km-long Fabry-Perot cavities in its arms. Virgo is located at the European Gravitational Observatory (EGO), close to Cascina (Pisa, Italy). It is the only ground-based antenna that, from its conception, aimed at reducing the detection band lower threshold down to 10 Hz. This goal has been achieved by isolating the interferometer mirrors from seismic noise, which, in the tens of Hz range, is many orders of magnitude larger than the small variation of the arm length induced by the gravitational wave passage.

EGO is responsible for the Virgo site, in charge of the construction, maintenance, and operation of the detector, as well as its upgrades. The goal of EGO is also to promote research and studies on gravitation in Europe. By December 2015, 19 laboratories plus EGO were members of the VIRGO Collaboration.

The construction of the initial Virgo detector was completed in June 2003, and several data-taking periods followed between 2007 and 2011. Some of these runs were done in coincidence with the two LIGO detectors. Then, a long upgrade to the second-generation detector, called Advanced VIRGO, started; its aim is to reach a sensitivity one order of magnitude better than the initial Virgo detector, allowing it to probe a volume of the Universe 1,000 times larger, making detections of gravitational waves more likely.

Advanced VIRGO started commissioning in 2016, joining the two Advanced LIGO detectors ("aLIGO") for a first "engineering" observing period in May and June 2017. Advanced VIRGO can detect gravitational waves with frequencies between 10 Hz and 10,000 Hz. These are the same frequencies as the sound waves that are audible to humans. Therefore, any signal measured by Advanced VIRGO can be sent to a loudspeaker, and we can hear the symphony of the Universe. This should allow the detection of gravitational radiation caused by the coalescence of binary systems (stars or black holes), pulsars, and those produced by supernovae in the Milky Way and in outer galaxies, for instance from the Virgo Cluster, hence the name of the project.

Virgo Interferometer

Virgo is a Michelson Laser Interferometer with two orthogonal arms, each 3 km long. For all practical purposes, Virgo works and operates on the same principle as that of LIGO. A beam splitter divides the incident laser beam into two equal components sent into the two arms of the interferometer. In each arm, two mirrors in a Fabry-Perot resonant cavity extend the optical length from three to about 100 km because of multiple reflections and therefore amplify the tiny distance variation caused by a gravitational wave. The two beams of laser light coming from the two arms are recombined out of phase on a detector so that, in principle, no light reaches the detector - a condition for destructive interference. The variation of the optical path's length, caused by the changing distance between the mirrors, produces a very small phase shift between the beams and, thus, a variation of the luminous intensity, which is proportional to the wave's amplitude.

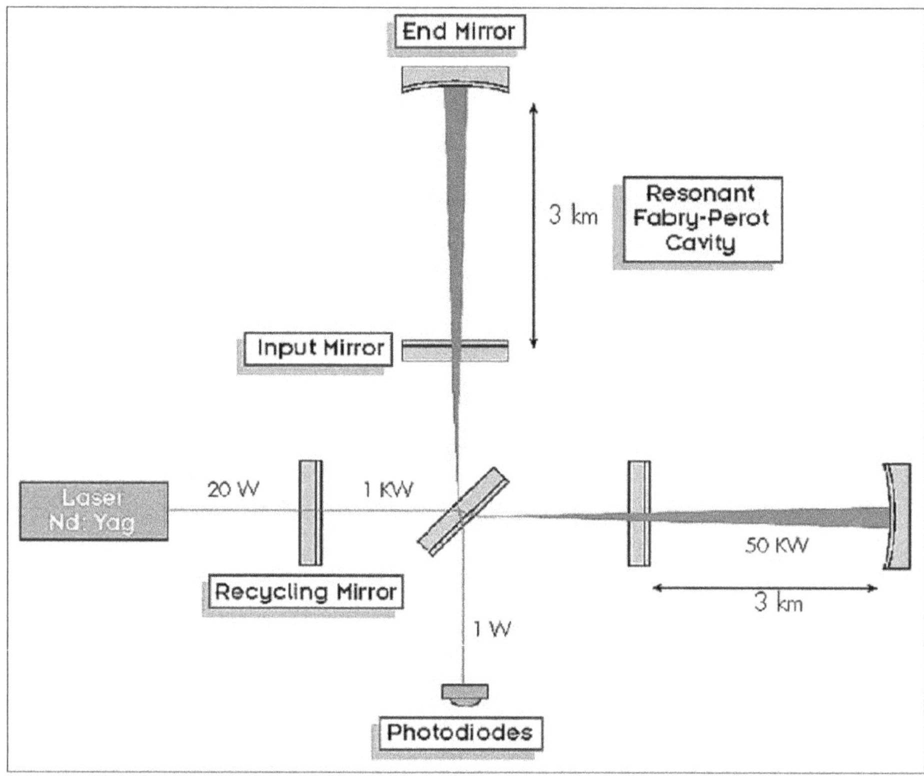

Figure 6-2.   Optical configuration of the first-generation Virgo detector. In the schematics, one can see the magnitude of the power stored in the various cavities (image credit: EGU)

However, in this scheme, a large fraction of the light is sent back towards the laser. In order to further increase the power, this light is sent back to the interferometer by a recycling mirror, in phase with the incident beam, thus increasing the light power that can reach several tens of kilowatts in the Fabry-Perot resonant cavities. High light power is important because it allows for the improvement of the sensitivity of the interferometer. With these resonant cavities coupled together, the interferometer can be seen as a giant light trap. If the optics were perfect and the mirrors perfectly stable, no light should normally reach the detector except when the interferometer plane is crossed by a gravitational wave. The quality and stability of the optics represent therefore one of the major challenges of the interferometer.

Virgo runs day and night, listening to all signals that arrive at any time from any part of the Universe. The data coming from the interferometer as well as the ancillary data necessary for its control (4 Mbytes/s) are subjected to a preliminary analysis for quality check and a quick detection of potentially interesting anomalous signals. The data is then made available to the scientific collaboration for a deeper analysis.

On 14 August 2017, LIGO and VIRGO detected a signal, GW170814, which was reported on 27 September 2017. It was the first binary black hole merger detected by both LIGO and VIRGO. Scientists also expect to observe several binary neutron star mergers that paved the way for the emergence of a new era - multi-messenger astronomy; as well as providing insights into binary evolution, nuclear physics, cosmology, and fundamental physics.

The 4th gravitational wave detection GW190521 from LIGO's and Virgo's 3rd observing run (O3) is very big news: the most massive merger of two black holes. The black hole collision formed a 142 solar mass black hole when the Universe was half its current age. This is the first direct observation of the birth of an intermediate-mass black hole. Another surprise is that the heavier black hole in the binary has 85 solar masses, which is not feasible according to our present understanding of stellar explosions. Scientists presume that this heavier black hole was born from an earlier binary merger. Scientists at the Max Planck Institute for Gravitational Physics in Potsdam, Hanover, and at Leibniz University, Hanover have made crucial contributions to this discovery and its interpretation.

This binary merger is very short and lasted only for about a tenth of a second. During this short time interval, its frequency chirps from 30 Hz to 80 Hz before the signal ends with the merger of the two black holes. As for the present understanding of star evolutions, black holes with either less than 65 solar masses or more than 120 solar masses will be formed but none in between because they do not explode as supernovae. The resulting gap in the black hole spectrum is called the "pair instability" gap. The nature and exact range of the mass gap is uncertain. Different estimates, models, and theories arrive at slightly different numbers. But it's there, and somehow, it's related to our overarching questions about black holes and how the Super Massive Black Holes (SMBHs) in galaxies got to be so massive.

Another unprecedented discovery has been unveiled by LIGO-Virgo scientists in August 2019. Data from the 3rd observation period (O3) of the Advanced LIGO and Advanced Virgo detectors reveal that at 21:10 (UTC) on the 14th of August 2019, the three instruments in the network detected a gravitational wave signal, called GW190814. The signal originated from the merger of an enigmatic couple: a binary system composed of a black hole, 23 times heavier than our Sun, and a much lighter object, about 2.6 times the mass of the Sun. The merger resulted in a final black hole about 25 times the mass of the Sun. It is this lighter object that makes GW190814 so special. It may just be either the lightest black hole or the heaviest neutron star ever discovered in a binary system. Another peculiar feature of GW190814 is the mass ratio of the objects in the binary system. The factor 9 ratio is even more extreme than was the case with the first detected merger of a binary with unequal masses, GW190412.

The world's three principal gravitational-wave detectors - LIGO in the US, VIRGO in Italy (VIRGO denotes the VIRGO Collaboration and the European Gravitational Observatory (EGO) Consortium), and now KAGRA in Japan - have signed a memorandum of agreement (MOA) on October 19, 2019, that covers scientific collaboration. The agreement includes the joint observation of gravitational waves and the sharing of data over the coming years, while it also foresees the expansion of the collaboration through the welcoming of new partners in the future.

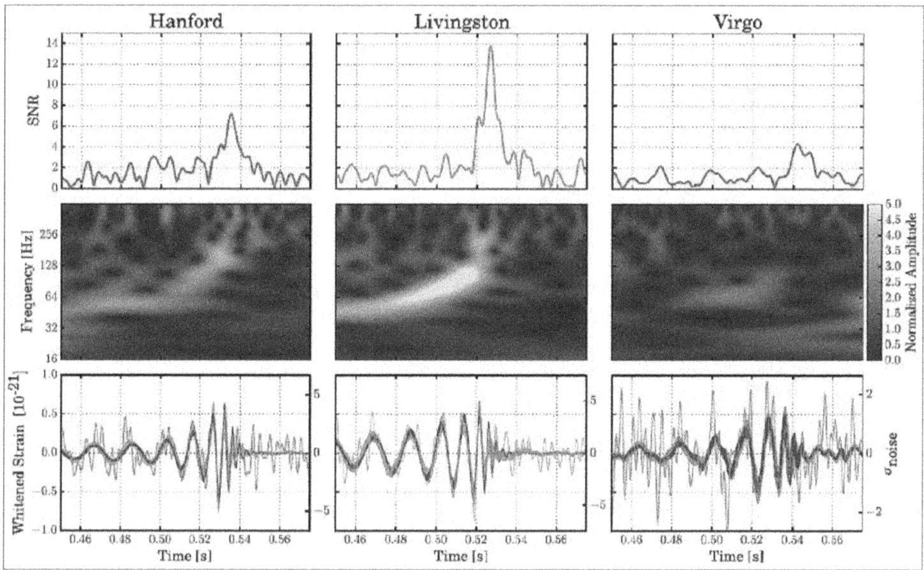

Figure 6-3. *The GW event GW170814 observed by LIGO Hanford, LIGO Livingston, and Virgo. Times are shown from August 14, 2017, 10:30:43 UTC. Top row: SNR (Signal-to-noise ratio is a measure used in science and engineering that compares the level of a desired signal to the level of background noise) time series produced in low latency and used by the low-latency localisation pipeline on August 14, 2017. The time series were produced by time-shifting the best-match template from the online analysis and computing the integrated SNR at each point in time. The single-detector SNRs in Hanford, Livingston, and Virgo are 7.3, 13.7, and 4.4, respectively. Second row: Time-frequency representation of the strain data around the time of GW170814. Bottom row: Time-domain detector data (in colour) and 90% confidence intervals for waveforms reconstructed from a morphology-independent wavelet analysis (light grey) and BBH (Binary Black Hole) models (image credit: LIGO and VIRGO Collaboration)*

VIRGO is highly useful in gravitational wave detection for several reasons:

- Design: VIRGO is specifically designed as a large interferometer to detect gravitational waves, with features such as long arms, suspended mirrors, and a vacuum-operated laser beam.

- Sensitivity: VIRGO is sensitive to a wide frequency range of gravitational waves, allowing it to detect various astrophysical events.

- Collaboration: VIRGO is part of scientific collaboration of multiple countries, enabling resource sharing and enhancing detection effectiveness.

- Validation: The initial VIRGO detector helped validate technical choices and improve the Advanced VIRGO detector.

- Increased Sensitivity: VIRGO, along with other detectors, significantly improves the sensitivity of the global network, increasing the chances of detecting faint signals.

- Localisation of Sources: The network, including VIRGO, can better localise the sources of gravitational waves, aiding follow-up observations by other telescopes.

- Confirmation of Signals: Multiple detector detections, including VIRGO, provide confirmation of gravitational wave signals, increasing confidence in observed events.

- Testing General Relativity: VIRGO's data, combined with other detectors, allows for testing Einstein's Theory of General Relativity.

- Multi-Messenger Astronomy: VIRGO's contribution to the network enables multi-messenger astronomy by combining gravitational wave observations with other types of astronomical observations.

Upgrades and Improvements: Virgo has undergone upgrades and improvements to enhance its sensitivity and performance. This includes the installation of a more powerful laser source, the replacement of steel wires with fused silica wires to suspend the mirrors, and the addition of a squeezed vacuum source to reduce noise.

- Collaboration with LIGO and KAGRA: Virgo has collaborated with the LIGO detectors in the US and the KAGRA detector in Japan. This collaboration has expanded the global network of gravitational wave detectors and improved the localisation and detection capabilities of the network.

- Data Sharing and Open Science: Virgo has made its data publicly available through the Gravitational Wave Open Science Centre. This allows scientists and researchers to access and analyse the data for scientific investigations and educational activities.

Overall, the experimental outcomes of Virgo have significantly contributed to the field of gravitational wave astronomy, providing valuable insights into the nature of the Universe and confirming the predictions of Einstein's Theory of General Relativity.

In the longer term, after accomplishing the primary goal of discovering gravitational waves, Virgo aims to be part of the birth of a new branch of astronomy by observing the Universe with a different and complementary perspective than current telescopes and detectors. Information brought by gravitational waves will be added to those provided by the study of the electromagnetic spectrum (microwaves, radio waves, infrared, the visible spectrum, ultraviolet, X-rays, and gamma rays), cosmic rays, and neutrinos. In order to correlate gravitational wave detection with visible and localised events in the sky, the LIGO and VIRGO collaborations have signed bilateral agreements with many teams operating telescopes to quickly inform (on the timescale of a few days or a few hours) these partners that a potential gravitational wave signal has been observed. These alerts must be sent before knowing whether the signal is real or not because the source (if it is real) may only remain visible for a short amount of time.

KAGRA

The Kamioka Gravitational Wave Detector (KAGRA) is a large interferometer designed to detect gravitational waves predicted by the General Theory of Relativity.ty. This Large-scale Cryogenic Gravitational Wave Telescope KAGRA is under construction underground in the Kamioka mine in Kamioka, Hida, Gifu Prefecture, Japan. The two unique features of KAGRA: It was constructed underground, and the test-mass mirrors are cooled to cryogenic temperatures

KAGRA is a Laser Interferometer with a baseline length of 3 km. When completed, it will be one of the most precise gravitational wave detectors in the world. Its aim is to build a worldwide gravitational wave detector network together with LIGO in the USA and Virgo in Europe. The construction of KAGRA is carried out by an international collaboration of many research institutes including The Institute for Cosmic Ray Research, the University of Tokyo as the main host, and NAOJ (National Astronomical Observatory of Japan) and the High Energy Accelerator Research Organization as co-hosts.

While the Advanced LIGO and VIRGO are considered second-generation detectors, these features make KAGRA a 2.5-generation detector, that is, the intermediate generation before 3rd-generation detectors, such as the Einstein Telescope (ET) and Cosmic Explorer (CE). The KAGRA experimental site is located under Mount Ikenoyama (elevation of 1369 m), in Gifu Prefecture, Japan.

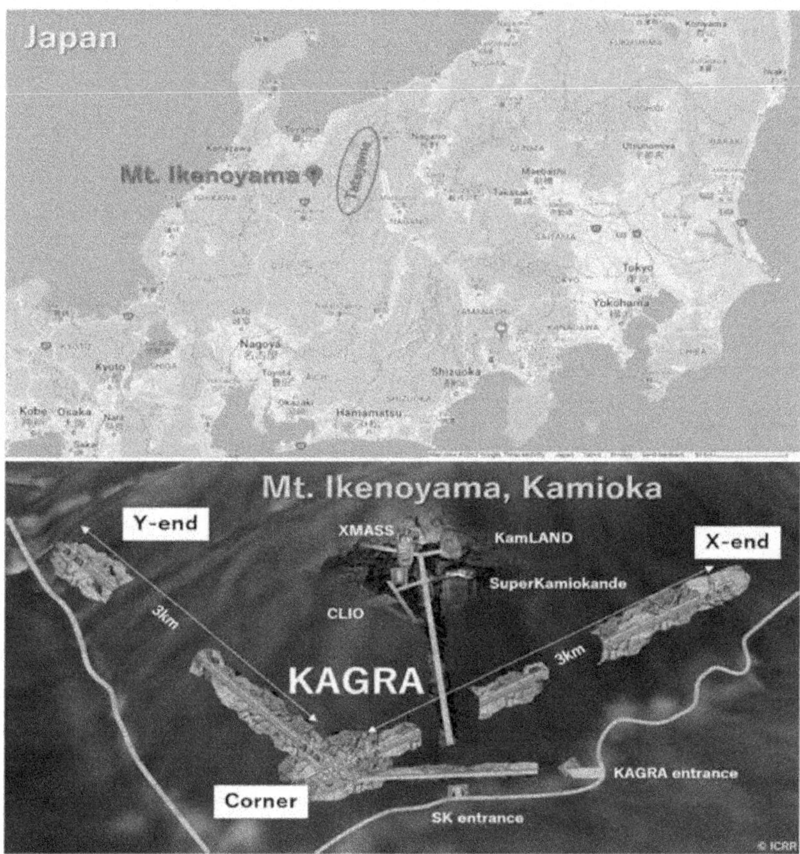

Figure 6-4. presents a schematic of the KAGRA experimental site. It consists of three stations (Corner, X-end, and Y-end), 2 arm tunnels (X-arm and Y-arm) 3 km in length, and several access tunnels. Under the same mountain, there are many neutrino experiments (Super-Kamiokande, KamLAND, and CANDLES), dark matter experiments, and other R&D experiments, including those of CLIO, which is the prototype of KAGRA.

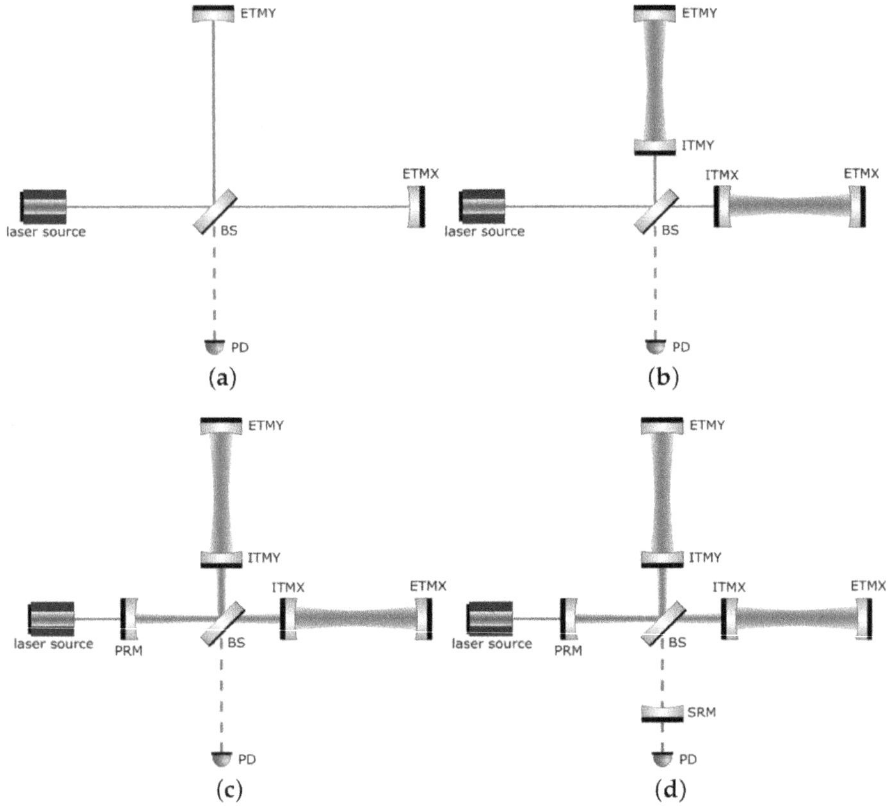

Figure 6-5. *Schematic image of the KAGRA interferometer configuration. A basic Michelson interferometer (**a**) consists of a laser source, beam splitter (BS), end test masses (ETMX and ETMY), and photodetector (PD). The Fabry-Perot arm cavities, which are composed of ETMs and input test masses (ITMX and ITMY), extend the effective arm length (**b**). The power recycling mirror (PRM) and signal recycling mirror (SRM) improve the sensitivity using the power recycling technique and resonant sideband extraction technique, respectively (**c, d**).*

(**a**) Michelson interferometer (MICH). (**b**) Fabry-Perot Michelson interferometer (FPMI). (**c**) Power recycling Fabry-Perot Michelson interferometer (PRFPMI). (**d**) Resonant sideband extraction (RSE) interferometer.

Because interferometers with longer arms are more sensitive to GWs, KAGRA has 3-km-long arms, like the other GW observatories. However, because a Michelson interferometer with an arm length of the kilometre order is not sufficient to achieve the required sensitivity for detecting a GW, current GW observatories combine multiple optical cavities with a Michelson interferometer to improve sensitivity.

For GW detection, all of the optical resonators must be held in resonance. To achieve this, the distance between the mirrors of the interferometer must be precisely controlled with an accuracy of a few hundred picometers. GW detection is possible only when the interferometer is maintained in a resonant state, which is referred to as "the interferometer is locked". The fundamental noises like seismic noise, mirror and suspension thermal noise, and quantum noise were reduced by designing KAGRA with two key features. One was the utilisation of an underground site to reduce seismic noise. The other was the utilisation of cryogenic mirrors to reduce thermal noise.

In April 2019, the Advanced LIGO detectors and the Virgo detector started their observation runs, which are called O3. In late March 2020, KAGRA was accepted to join O3 as a member of the scientific collaboration with LIGO and VIRGO. However, LIGO and VIRGO stopped their operations in March 2020 because of the COVID-19 pandemic. In this situation, GEO 600, which is the interferometric GW observatory in Germany with an arm length of 600 m, continued the observation, and GEO 600 and KAGRA started a joint observation run called O3GK. This was the first international observation run and a significant milestone for KAGRA.

After three years of downtime for crucial upgrades, the world's gravitational-wave observatories LIGO and VIRGO were supposed to restart in 2023, more sensitive than ever to detect the tiniest deformations in spacetime. But technical problems scuttled that optimistic scenario, and expectations for the 20-month campaign, known as Observing Run

4 (O4) with desired consistency, have fallen behind schedule. Italy's Virgo detector is undergoing repairs to fix damage caused by broken glass fibre. Additionally, Japan's new KAGRA is far from its intended sensitivity and will be ready for restart only in late 2024, after shutting down for troubleshooting for a few months.

This leaves only the LIGO detector to emerge fully functioning from the shutdown. LIGO's upgrade alone should ensure the project discovers nearly three hundred new events in O4—roughly one every other day.

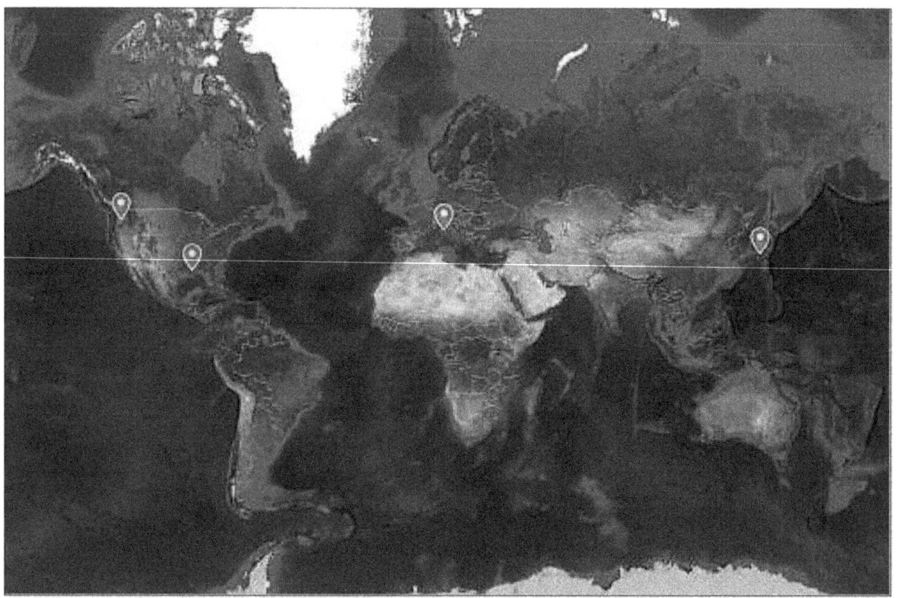

Figure 6-6. Location of the three gravitational-wave detectors, LIGO, Virgo, and KAGRA, on a world map (image credit: LIGO, Virgo & KAGRA)

LIGO and VIRGO are so sensitive, in fact, that they must account for the noise produced by the impacts of photons from the laser. "When the photons reflect, they provide a momentum transfer to the mirror." To limit this noise, the photons must all have the same phase; essentially, they need to all hit the mirror at about the same time. Previously, LIGO and VIRGO had only been able to "squeeze" light into the same phase at high frequencies above 50 Hz. Thanks to upgrades, which included installing a 300m tunnel at the LIGO sites, the detector can now squeeze light down to 30 Hz. This low-frequency squeezing has dramatically improved LIGO's sensitivity, which is measured as the distance to the farthest detectable binary neutron star/black hole merger.

LIGO was previously sensitive out to 330 million light-years. It has now improved its range by about 50 per cent to gauge the cosmic depths of about 500 million light-years. At Virgo, upgrades were going on well until November 2022, when researchers installed a noise mitigation system on a mirror. It was soon discovered that one of the delicate glass fibres suspending the mirror had snapped. A similar incident happened in 2017 when a glass fibre broke, causing inordinate delays. For both mishaps, Virgo researchers identified a likely common cause: during upgrades, particles of dust settled on the fibres and weakened them. The Virgo upgrade process has been ongoing since 2019.

Virgo is not the only detector to struggle. Three years ago in 2020, experts optimistically predicted that KAGRA would be sensitive to about 424 million light-years when it starts its operations. Its first-of-its-kind cryogenic design and underground location were supposed to offer extra protection against environmental noise. But various setbacks have held KAGRA to less than one per cent of the planned sensitivity, extending to only three million light-years, which limits the detection of gravitational waves within or just outside of our own galaxy.

After upgrading and modifications, KAGRA's sensitivity may be boosted further by a factor of 10—a value that would still fall short of its original goal. What is hampering the detector is the formation of fog on the cryogenic mirrors, which necessitated the development of new mirrors down to a frigid 2.0 K without forming frost. The 2024 Noto earthquake on 1 January 2024, whose epicentre was about 120 km from KAGRA, damaged its mirror suspension mechanism. As of 5 February 2024, the project is expected to return to observation in January 2025.

GW detection with KAGRA is vital for promoting GW astronomy and will also serve as the basis for introducing new technologies in future GW observatories.

The LIGO-Virgo-KAGRA international network of gravitational wave observatories announced the discovery of an exceptional astronomical event. The event is named GW230529, as it was detected during the first week of the current observing run on May 29, 2023, by the LIGO Livingston Observatory in the US The signal resulted from the collision between a neutron star and another compact object of undetermined nature, which lies in the "mass-gap" between the heaviest known neutron stars and the lightest black holes. The gravitational wave signal alone cannot reveal the nature of this object. Future detections of similar events, especially those accompanied by bursts of electromagnetic radiation, could hold the key to solving this cosmic mystery.

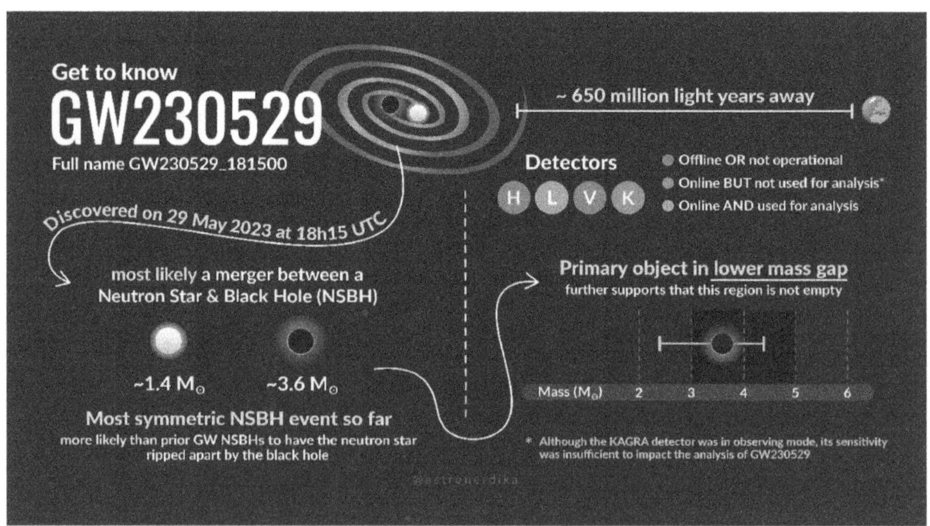

Figure 6-7.   Analysis of GW230529 NS-BH merger

GW230529 is not the first compact object to be found in the "mass-gap". In the previous observing run, the LVK collaboration also found the event GW190814, reported as a collision between a massive black hole and another unknown compact object in the mass gap.

Of all the neutron star-black hole mergers observed to date, GW230529 has the least difference between the masses of the colliding objects. This observation suggests that future observations of similar systems may be more common than researchers had previously thought. Systems like GW230529 have a greater chance of producing an electromagnetic counterpart as the black hole is not massive enough to swallow the neutron star before it is ripped apart, so although this particular event was observed only in gravitational waves, it increases the expectation that more such events will also be observed with electromagnetic waves in the future. It is for the precise localisation of such events that LIGO-India is going to play an important role once it is operational.

CHAPTER – 7

Indian Collaboration

More than a thousand scientists from 15 countries have participated in the discovery of GW150914. Led by senior scientists like Bala Iyer at the Raman Research Institute, Bangalore, and Sanjeev V. Dhurandhar at the Inter-University Centre for Astronomy and Astrophysics (IUCAA) Pune, 35 scientists from nine Indian institutions have made direct contributions to the discovery of gravitational waves at LIGO. The GW150914 discovery paper has authors from these Indian institutions. This programme ranged from designing algorithms that were used to analyse whether the signals registered by the detector were indeed from a gravitational wave, to working out parameters such as establishing the energy and power radiated during the merger, orbital eccentricity, estimating the mass and spin of the final black hole—the Indian scientists' contribution is enormous. Some of this work was carried out on high-performance computing facilities at IUCAA, Pune, and International Centre for Theoretical Sciences (ICTS), Bengaluru.

Physicists at the Inter-University Centre for Astronomy and Astrophysics (IUCAA), Pune, an autonomous institution of the University Grants Commission, India, made significant contributions to this discovery. They are Professors Sukanta Bose, Sanjeev Dhurandhar (Emeritus), Sanjit Mitra, Tarun Souradeep, postdoctoral fellow Anuradha Gupta, PhD students Anirban Ain, Bhooshan Gadre, Nikhil Mukund, and visiting scientists Jayanti Prasad and Sharad Gaonkar. They are all active members of the LIGO Scientific Collaboration (LSC).

IUCAA's involvement in the discovery spanned across a wide range of areas of this multidisciplinary search, beginning with the basic idea used in finding the weak and short-lived gravitational wave signal in the very noisy data of the LIGO detectors. That idea, due to Dhurandhar and B. Sathyaprakash (then a postdoctoral fellow at IUCAA, now in Cardiff) from 1991, was of matching thousands of wave patterns expected from black hole binaries, as derived from Einstein's theory, with the data from the detectors. This maiden detection relies on this fundamental work.

Some other Indian astronomers were also involved in searching for possible electromagnetic counterparts using optical telescopes, while some other theoretical cosmologists computed whether the observed signal agreed with the predictions of Einstein's General Theory of Relativity.

LIGO-INDIA

The first direct observation of gravitational waves is a milestone, showing a new way to explore the distant non-thermal Universe. Several new detectors are underway: One of them is LIGO-India.

LIGO-India is a planned advanced gravitational-wave observatory envisaged as a collaboration between several Indian institutions, the LIGO laboratories in the US, and LIGO's international partners in Australia, Germany, and the UK. The project will relocate a third LIGO interferometer, identical in design to the other two and originally planned for installation at Hanford. Both KAGRA (of Japan) and LIGO-India will operate as part of a gravitational detector network together with LIGO and VIRGO, contributing to superior source localisation.

LIGO-India is an initiative to set up advanced experimental facilities for a multi-institutional observatory project in gravitational-wave astronomy to be located near Aundha Nagnath, Hingoli District, Maharashtra, India. Scientists travelled to different cities and scientific institutions, sometimes by road and sometimes by air, in the quest to find the ultimate place to build LIGO in India. In the absence of project funds prior to 2016, the team carried out seismic studies by borrowing instruments from geophysics colleagues and training people to use them. Also using data from the Indian Space Research Organisation (ISRO), scientists considered about 40 options across the Indian subcontinent. Finally, the scientists' group zeroed in on Aundha, which is far from the crashing waves along the Indian coastline and the rumbling of trains along India's railway systems. The site is situated on a natural barren plateau of the Deccan Traps, a volcanic formation in an otherwise fertile—and populated—region. One of the group's scientists, Tarun Souradeep, says, "It took five years to find the perfect location. We have a technically excellent site where we didn't displace anyone. We are very proud of that fact."

Figure 7-1.   Schematic architectural design of Proposed LIGO-India.

In April 2023, the Indian government approved construction, and by the end of the decade, they expect to see the observatory's first detections of gravitational waves. LIGO-India will be the second gravitational-wave detector in Asia, joining the previously built KAGRA detector in

Japan. Detecting a gravitational wave from multiple locations on the globe allows scientists to triangulate where it's coming from. "Being like the two LIGO detectors but much farther away, LIGO-India has the potential to have comparable sensitivity and hence improve sky localisation of gravitational-wave sources," says Bala Iyer, a theoretical physicist, among the project leads for LIGO-India as part of the IndIGO consortium. (https://www.gwindigo.org/tiki-index.php).

Given its position on the opposite side of the globe from the United States, the Indian LIGO detector will be misaligned with the Washington and Louisiana detectors—and this is another key to its usefulness. Because of its location, LIGO-India will measure gravitational waves in a different polarisation. "It's very similar to polarised sunglasses," says Adhikari, a professor of physics at Caltech in the US and an associated faculty member at the International Centre for Theoretical Sciences in Bengaluru, India.

Binary star systems, for example, emit gravitational waves that are polarised in a circular pattern. The US LIGO detectors sense only one part of this polarisation, and scientists cannot ascertain whether they are seeing the orbit "face on" or "edge on." This missing information makes it harder to estimate the distance to the source of the gravitational waves. LIGO-India will make it easier to find events like neutron star mergers in the sky.

LIGO-India is a collaboration between the LIGO Laboratory (operated by Caltech and MIT) and three institutes in India:

1. The Raja Ramanna Centre for Advanced Technology (RRCAT, in Indore),

2. The Institute for Plasma Research (IPR) in Ahmedabad,

3. The Inter-University Centre for Astronomy and Astrophysics (IUCAA) in Pune.

Extracting the information carried by the gravitational waves to address questions in both physics and astronomy depends on our ability to identify where the individual sources are in the sky. This requires a network of detectors spread widely over the Earth. The two detectors of LIGO in the United States along with the VIRGO detector in Italy can triangulate gravitational wave sources over part of the sky. LIGO-India will enable scientists to locate sources across the entire sky. This dramatic improvement with LIGO-India would come in the ability to localise GW sources in the sky, due to its geographic advantage—this detector will be far from the countries with existing detectors.

Maps of the sky comparing how accurately the positions of sources can be located without and with LIGO-India. (Credit: Stephen Fairhurst, "Improved source localisation with LIGO-India")

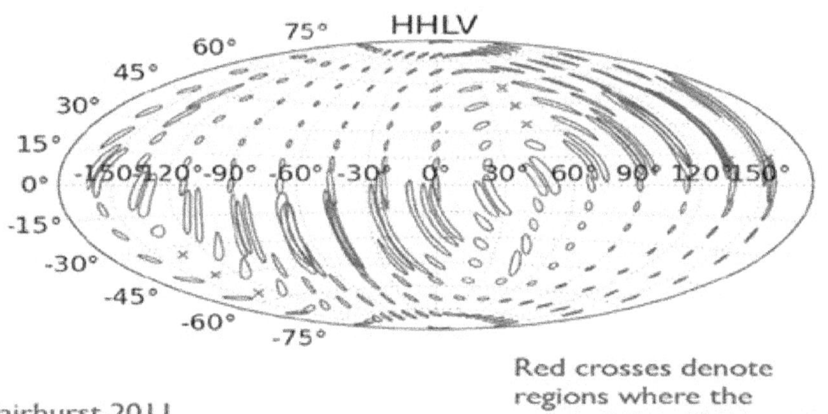

Original Plan: 2 LIGO + 1 VIRGO

The ellipses on the sky maps below show how much more accurate sources can be found with LIGO-India. The dramatic improvement with LIGO-India is the key scientific motivation for this project.

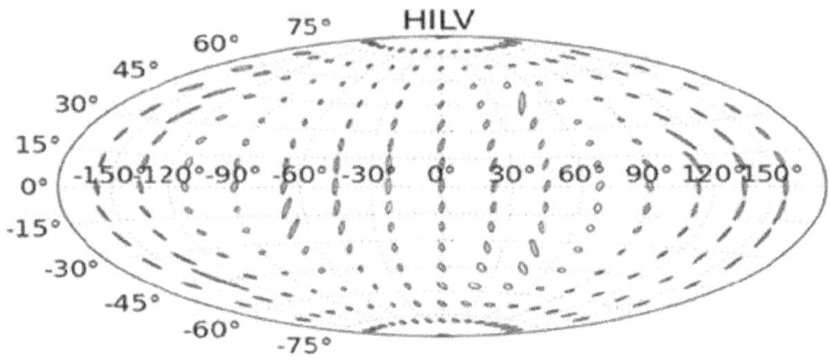

Fairhurst 2011

LIGO-India Plan: Indigo + LIGO + VIRGO

With the two present detectors of LIGO based in the US, each with two 4-km-long detector 'arms', and the Virgo with 3-km-long detector 'arms' based in Italy, a detector in India will provide a baseline almost the diameter of Earth. Because we are placed on an almost opposite side of Earth from the US detector, for example, we get a baseline of 39 milliseconds [multiplied by the speed of light], whereas the diameter of Earth is 41 or 42 milliseconds. So, it is close to Earth's diameter. But with the 3rd detector, the localisation of the source will be much improved. Also, the sky coverage will be better because, in India, we will have a different orientation. While researchers were able to localise the first detection to 600 square degrees with the two LIGO detectors— with a 3rd detector in India operating, the same source would have been localised to between 2 and 7 degrees: an improvement of a factor of 100. (For reference, a full Moon is about 0.2 square degrees in the sky). This reduces the area that astronomers need to search through to pinpoint the merger events. In actuality, with LIGO-India now approved for construction, there will be five gravitational-wave detectors working together in one network, further improving the localisation of mergers: LIGO-India, LIGO Hanford, LIGO Livingston, Virgo, and KAGRA.

Interferometric gravitational wave detectors incorporate cutting-edge technology from various fields. Glimpses of a few such technological marvels that will be used in LIGO-India are provided below.

Laser

The LIGO interferometer hosts a laser at the 1064 nm target wavelength. A non-planar ring oscillator (NPRO) generates a 2 W 'seed' beam that is amplified through several stages until it can reach the desired 200 W at the output. The entire system is called the pre-stabilised laser system (PSL).

Optics

The LIGO core optics or test masses are currently made of ultra-high purity fused silica and coated with titania-doped tantala so that they have minimal absorption at the working wavelength of 1064nm. The mirror coatings are also dichroic in that they provide some reflectance at (1064/2) nm = 532 nm, which is used for auxiliary measurements. The mirrors are specified to have absorption in the order of 3ppm and a figure error in the order of 0.35nm. This poses a major challenge for materials research given the size of the mirrors at 34cm in diameter and weighing 40kg.

Suspensions and Vibration Isolation Systems

LIGO vibration isolation can be broadly classified as active damping and passive damping systems that together ensure that the detector components are isolated from any external vibration. Active damping systems probe the environment for vibrations at different frequencies using sensors and generate counter-movements, thereby isolating the instrument from such noise. Passive damping systems include the seismic isolation and suspension sub-systems for the optics in LIGO.

Ultra-high Vacuum (UHV)

LIGO's vacuum system is one of the largest sustained vacuum systems and also one of its kind. The detector is maintained at one-trillionth of the atmospheric pressure at sea level. Such ultra-high vacuum becomes necessary to eliminate noise in length changes of the interferometer arm brought by any residual air in the chambers affecting the detection of a passing gravitational wave and also to avoid absorption of the laser light by any residual dust or gas molecules, thereby causing unwanted scattering.

Beam Tube

The beam tube is one of the important parts of the entire vacuum system. It is a cylindrical structure with a length of 4 km in each perpendicular direction.

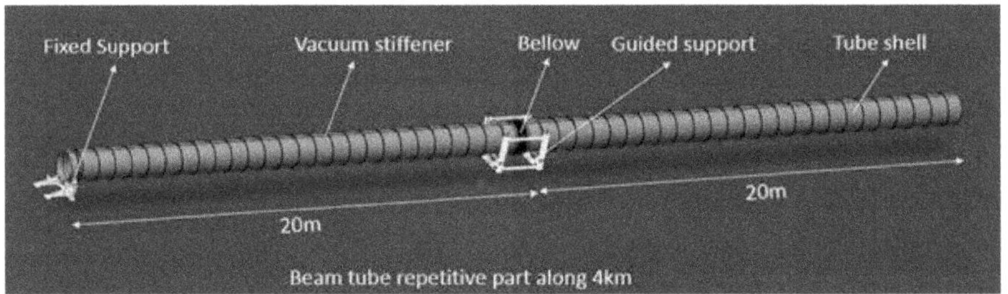

The LIGO-India project is a mega-science project in astronomy on Indian soil. It promises breakthrough research outcomes, the development of cutting-edge technology, and opportunities for students and researchers. The project will benefit India in other ways as well. The first detection of gravitational waves was one of the highest-profile scientific discoveries of our time. Engaging the Indian scientific community in this quest will raise the visibility and appeal of experimental science in India. The presence of a world-leading facility in India can be used to attract students and inspire them to pursue technical careers.

A Grand Collaborative Effort

- The LIGO Laboratory will provide the hardware for a complete LIGO interferometer, technical data on its design, installation, and commissioning, training and assistance with installation and commissioning, and the requirements and designs for the necessary infrastructure (including the vacuum system). The components for the LIGO-India detector have already been fabricated as part of the Advanced LIGO project, funded by the National Science Foundation (NSF).

- India will provide the site, the vacuum system, and other infrastructure required to house and operate the interferometer, and all labour, materials, and supplies for installation, commissioning, and operations. Funding for the LIGO-India facilities will come from the Department of Atomic Energy (DAE) and the Department of Science and Technology (DST), with DAE acting as the lead agency.

The site selection committee at IUCAA has looked at over 22 sites and worked to shortlist a few of them based on site selection criteria like low 'seismicity', low human-generated noise, socio-economic considerations, air and road connectivity, and data and telecommunications accessibility. The IPR team has prepared system requirement documents, conceptual drawings, and engineering drawings for the sophisticated civil infrastructure and ultra-high vacuum systems in consultation with LIGO labs.

Once it becomes operational, LIGO-India will be scientifically managed and operated in collaboration with the US LIGO detectors to optimise the scientific return.

LIGO-India's Outcome Programmes

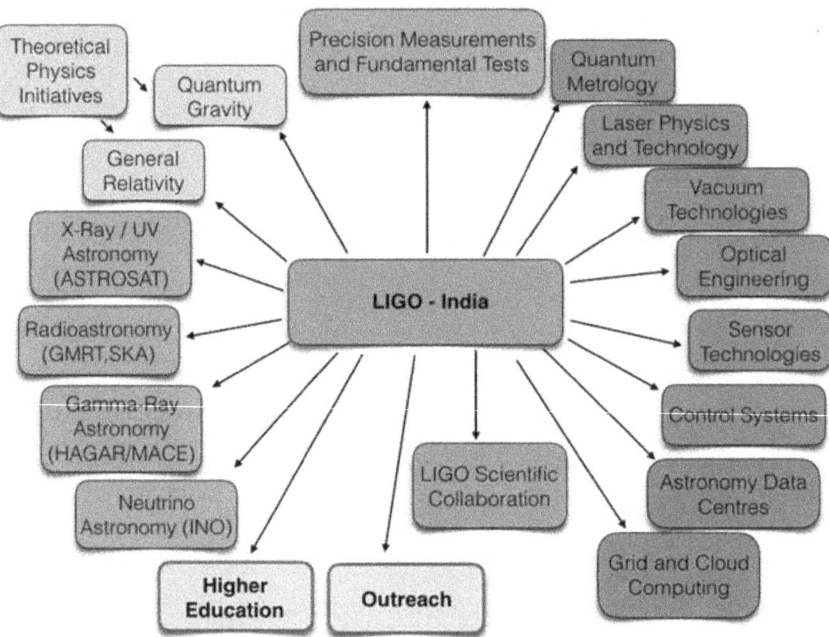

Education and Public Outreach

The LIGO-India project has an Education and Public Outreach (EPO) team to go hand in hand with the progress of the project itself.

Such an effort is needed to:

- To create interest among science and engineering students towards gravitational wave science.

- Maintain continued interest in the project and create awareness about the project among the general public.

- Conducting various events for the members of the local community as well as with governmental and service organisations.

Outreach near the site.

1. GW science and astronomy workshops for schools and colleges

2. Stargazing programmes for the public

3. Forming local amateur astronomy clubs and conducting club activities

4. GW talks in local colleges: Reaching colleges/universities across India, conducting GW Talks and Workshops in various colleges. Active participation in national-level college tech-fests

LIGO-India Scientific Collaboration (LISC)

Pan-Indian participation in the LIGO Scientific Collaboration (LSC) is conducted through the LIGO-India Scientific Collaboration (LISC). This includes activities related to the LIGO-India observatory and participation in the various working groups of the LSC. An Executive Committee has been set up to manage the LISC. Its present members are Prof Bala Iyer (ICTS-TIFR), Prof Sanjit Mitra (IUCAA), Prof Archana Pai (IIT Bombay), and Prof Anil Prabhakar (IIT Madras). The current PI/Co-PI of LISC are Prof Archana Pai and Prof Bala Iyer. All policy decisions of LISC will be made by the LISC Council. The Council comprises all senior members from Indian groups at different institutions.

The Indian Initiative in Gravitational-wave Observations (IndIGO) consortium has been playing this pivotal role since 2012.

The premier institutions of India in collaboration with LISC:

1. Chennai Mathematical Institute (CMI), Chennai

2. Directorate of Construction, Services & Estate Management (DCSE) Mumbai

3. Indian Institute of Science Education and Research (IISER), Kolkata

4. Indian Institute of Science Education and Research (IISER), Pune

5. Indian Institute of Technology (IIT) Bombay, Mumbai

6. Indian Institute of Technology (IIT), Gandhinagar

7. Indian Institute of Technology (IIT), Hyderabad

8. Indian Institute of Technology (IIT) Madras, Chennai

9. Institute for Plasma Research (IPR), Gandhinagar, Gujarat

10. International Centre for Theoretical Sciences, TIFR (ICTS-TIFR), Bengaluru

11. Inter-University Centre for Astronomy & Astrophysics (IUCAA), Pune

12. Raja Ramanna Centre for Advanced Technology (RRCAT), Indore

13. Tata Institute of Fundamental Research (TIFR), Mumbai

14. Saha Institute of Nuclear Physics (SINP), Kolkata

15. CSIR-Central Glass & Ceramic Research Institute (CGCRI), Kolkata

16. Indian Institute of Technology, Calicut

LIGO-India will play a key role in locating and deciphering GW sources with implications for astrophysics, cosmology, and eventually fundamental physics.

Some of the prominent Indian scientists who collaborated with LIGO laboratories in the discovery of gravitational waves are:

Bala Iyer, International Centre for Theoretical Sciences (ICTS), TIFR, Mumbai.

Prof Bala Iyer is the Chairperson of the Indian Initiative in Gravitational-Wave Observation (IndIGO) consortium, which is the proposer of LIGO-India, and he is one of its principal leads. Prof Bala Iyer is one of the pioneers in the modelling of high-accuracy gravitational waveforms from the inspiral of Neutron stars and Black holes. Before joining as a Visiting Professor at ICTS, Bengaluru, he held various academic positions at Raman Research Institute, Bengaluru. He is a Fellow of the American Physical Society and the International Society of General Relativity and Gravitation.

Prof Bala Iyer's principal research interest over the last two decades relates to the computation of high-accuracy gravitational waveforms for inspiralling compact binaries (ICB) of neutron stars and black holes using multipolar post-Minkowskian methods. These high-accuracy waveforms pave the way for the construction of gravitational wave templates used in the gravitational wave (GW) data analysis channels of detectors like LIGO and VIRGO. They also lead to a more accurate parameter estimation accuracy, which would be crucial in extracting astrophysical information from terrestrial GW detectors like LIGO and cosmological information from space-based detectors like Laser Interferometer Space Antenna (LISA).

Over the last 15 years, he has been involved in the Research Education Advancement Programme (REAP) for undergraduate teaching in physics at the Jawaharlal Nehru Planetarium, Bangalore. Prof Iyer was elected as a Fellow of the American Physical Society (2012) and the International Society of General Relativity and Gravitation (2013).

Prof Tarun Souradeep, Director and Professor, Raman Research Institute

One of the co-proposers and investigators of LIGO-India, which includes the construction and operation of a gravitational wave observatory on Indian soil.

Professor Tarun Souradeep assumed the role of Director at the Raman Research Institute (RRI) on the 20th January 2022. Before joining RRI, he was a professor and the Chair of the Physics Department at IISER, Pune. Prof Souradeep's expertise lies in Cosmology, Gravitational Wave Physics, and Astronomy, fields in which he has made significant contributions and successfully spearheaded a robust Indian initiative addressing issues on par with international research standards. The impact of the Indian team under his leadership is widely acknowledged, particularly in the field of cosmology through cutting-edge cosmic microwave background experiments (CMB) and the emerging field of gravitational wave (GW) astronomy.

He is the spokesperson (Science) for LIGO-India and Member Secretary of the LIGO-India Scientific Management Board. Together with Prof Bala Iyer (then at RRI), he was one of the lead proposers of the national mega-science LIGO-India project way back in 2011. He is a recipient of various awards including the DST Swarnajayanti Fellowship, and Gruber Cosmology Prize 2016 and 2018. He was awarded the Special Breakthrough Prize* in Fundamental Physics 2016, Fundamental Physics Prize Foundation, 2016 shared with LIGO Scientific Collaboration. He is a fellow of the National Academy of Sciences and the Indian Academy of Sciences. His election to the fellowship of the International Society on General Relativity and Gravitation is a testament to his standing among his international peers.

(The Special Breakthrough Prize comes with two and a half times the amount of the Nobel Prize. It was instituted by Yuri Milner (Russian Industrialist) and is given to scientists who have made a significant breakthrough. The whole amount of 3 million dollars is distributed among the team of scientists who have made the discovery of GW 150914. One million dollars is for the LIGO founders Ronald W. P. Drever, Kip S. Thorne, and Rainer Weiss. The remaining 2 million is being given to over 1000 collaborating scientists. There are about 37 of them from India—Prof Souradeep is one among them—and eight are from IUCAA. Several of them are PhD students).

Sanjeev Dhurandhar

Prof Sanjeev Dhurandhar is a physicist and Emeritus Professor at the Inter-University Centre for Astronomy and Astrophysics (IUCAA), Pune. He has made seminal contributions to gravitational-wave physics, particularly to the data analysis techniques for the detection of gravitational waves. Presently, he also serves as the Science Adviser to the Indian Initiative in Gravitational-Wave Observations (IndIGO) Consortium Council.

He is not only a Doyen of Gravitational Wave Physics in India but also one of the giants of gravitational wave physics research worldwide. He has been leading the Indian research and experimental efforts in the area of gravitational waves for three decades. His research interest is the detection and observation of Gravitational waves, their data analysis, and modelling of gravitational wave detectors. Prof Dhurandhar was part of the Indian team which contributed to the detection of gravitational waves.

He worked mainly on the extraction of the gravitational signal from the noise. There is a vast volume of data being received, and the gravitational signal itself is weak, so sophisticated mathematical and statistical techniques are designed to extract the signal from noise. In 1991, along with his then-colleague BS Sathyaprakash, Dhurandhar set up a basic method for the extraction of the inspiraling binary signal from detector data. Some of the mathematical techniques have stood the test of time.

He is the science adviser to the IndIGO consortium Council. He has been leading the Indian research and experimental efforts in the area for the last three decades. Prof Dhurandhar works at the Inter-University Centre for Astronomy and Astrophysics (IUCAA), Pune, on problems in General Relativity and black hole thermodynamics. He played an important role in the first direct discovery of gravitational waves, GW150914, by the Laser Interferometer Gravitational-wave Observatories (LIGO), as well as helped ensure India will get a LIGO of its own in the mid-2020s.

Professor CV Vishveshwara, Indian Institute of Astrophysics (IIA).

Prof Vishveswara has done pioneering work on the problem of perturbed black holes and discovered that the perturbed black hole settles down to equilibrium by emitting gravitational waves. His seminal work includes perturbing the black hole with a small energy field. He first discovered that the black hole, when disturbed, rings pretty much like a bell by emitting gravitational waves. These characteristic modes of black holes are called quasinormal modes (QNM). The QNM signals are one of the unambiguous signatures of black holes. The first detection of gravitational waves from a black hole merger, GW150914, is a significant step in the direct identification of QNM signals. His own work was highlighted in the discovery papers of gravitational waves, and rightfully so. Unfortunately, he fell ill soon after the announcement of the detection of GW150914 and passed away in January 2017.

Other young Indian scientists like Rajesh Nayak at IISER Kolkata, Archana Pai at IISER Trivandrum, Bala Iyer's student Arun K.G. at the Mathematical Institute in Chennai, R.R. Sengupta at IIT Gandhinagar have done a lot of work in this field, and it is a growing community. "It's heartening to see the involvement of many young scientists in India in this discovery," says LIGO member BS Satya Prakash from Cardiff University in the UK.

One of the research outcomes of one of these scientists is the recent paper published by IIT-H (Hyderabad) researchers in July 2023. IIT-H researchers, as part of the Indian Pulsar Timing Array (InPTA) consortium, have found evidence for ultra-low frequency gravitational waves originating from a large number of 'dancing monster black hole pairs' more than a million times the size of the Sun. The researchers, part of an international team of astronomers from India, Japan, and Europe, have published results from monitoring 'pulsars', nature's best clocks, using six of the world's most sensitive radio telescopes, including the country's largest telescope uGMRT. (PTA - Pulsar Timing Array: see in Appendix-I)

These results provide a hint of evidence for the relentless vibrations of the fabric of the Universe and are a crucial milestone in opening a new, astrophysically rich window in the gravitational wave spectrum. Such 'dancing monster black hole pairs', expected to lurk in the centres of colliding galaxies, create ripples in the fabric of the cosmos, and astronomers call them 'nanohertz' gravitational waves as their wavelengths can be many lakhs of crores of kilometres.

The relentless cacophony of gravitational waves from a large number of supermassive black hole pairs creates a persistent humming in our Universe. The team, consisting of members of the European Pulsar Timing Array (EPTA) and Indian Pulsar Timing Array (InPTA) consortia, published their results in two papers in the Astronomy and Astrophysics journal. These results include an analysis of pulsar data collected over 25 years with six of the world's largest radio telescopes. IIT-H has been part of InPTA since 2018, and some of the past InPTA students are pursuing higher studies in astrophysics and related industries. "I am delighted that the state-of-the-art 'NSM Param Seva' computing facility at IIT-H has helped to create these groundbreaking results. This achievement also underscores the power of collaboration in attaining scientific benchmarking results," said IIT-H director Prof BS Murthy. The InPTA experiment involves researchers from NCRA (Pune), TIFR (Mumbai), IIT (Roorkee), IISER (Bhopal), IIT (Hyderabad), IMSc (Chennai), and RRI (Bengaluru) along with colleagues from Kumamoto University, Japan.

A Multifaceted Opportunity

LIGO-India will not only provide an advantage to the international gravitational wave community and bring India to the vanguard of gravity research, but it also has multidisciplinary benefits for India's astrophysics research, high-end technological development, and human resource development in general. Since most of the components will be made in India, it will improve the technological expertise of Indian scientists and engineers. Spin-offs to other science, engineering, and industrial applications will provide unprecedented commercial opportunities for Indian industries and businesses in collaboration with research institutes and universities. LIGO-India will be transformational for Indian physics, technology, and astronomy sciences and drive accelerated growth in these areas over decades to come.

LIGO-INDIA Timeline

2012–2013 Site survey, validation, selection, and acquisition

2014–2015 Site preparation, Design, and Drawings for Buildings Tendering for Civil infrastructure, construction of Central and End stations.

2012 – 2015 On-site training and participation at LIGO-USA during the initial phases of LIGO-USA assembly and tests.

2016 March Indo-US Memorandum of Understanding (MOU) between the Department of Atomic Energy and the Department of Science and Technology, India, and the National Science Foundation, USA, was signed on March 31, 2016, in Washington DC in the presence of the Honourable Prime Minister of India, Sri Narendra Modi.

In October 2016, the state cabinet approved government-held land for LIGO-India on Oct. 4, 2016.

July 2017: Entire funds that will be committed before the full DPR funding is sanctioned.

In August 2017, state government land was formally transferred to LIGO-India.

In July 2018, the DAE sets in place the necessary project appraisal committee for DPR.

2018 August First LIGO-India Apex Committee meeting with DAE and DST top brass.

In March 2016, a month following LIGO's historic announcement of the first detection of gravitational waves, NSF and LIGO officials met with the Indian Prime Minister to sign a "Memorandum of Understanding" (MOU), representing India's commitment to building a 3rd LIGO detector in India.

Left to right: Dr Rana Adhikari (Caltech), Karan Jani (Georgia Tech), Nancy Aggarwal (MIT), the PM of India, Dr France Córdova (Director, NSF), Dr Dave Reitze (LIGO Executive Director, Caltech), Dr Rebecca Keiser (Head, NSF, Office of International Science and Engineering), Fleming Crim (NSF, Assistant Director for Mathematical and Physical Sciences).

The LIGO-India project has an Education and Public Outreach (EPO) team to go hand in hand with the progress of the project itself. The programme is dedicated to making the concepts of gravitational waves and LIGO accessible to everyone, from the general public and children to STEM graduates seeking to expand their knowledge and skills.

CHAPTER – 8

Roadmap

The first direct gravitational wave detections have undoubtedly opened a new window into the Universe. The scientific insights emerging from these detections have already revolutionised multiple domains of physics and astrophysics, yet they are 'the tip of the iceberg', representing only a small fraction of the future potential of GW astronomy.

THIRD GENERATION GW DETECTORS

The 3rd generation of gravitational wave detectors refers to a new generation of advanced interferometers that are currently being proposed and developed. These detectors aim to significantly improve upon the sensitivity and capabilities of the current second-generation detectors like LIGO and VIRGO. Some of the proposed 3rd-generation gravitational wave detectors include the Einstein Telescope (ET), Cosmic Explorer (CE), and LISA (Laser Interferometer Space Antenna).

It has been realised that every gravitational wave that passes through Earth comes in with a specific orientation, and only the orientations that cause substantial shifts in both perpendicular laser arms of an individual detector can lead to detection. The twin LIGO Hanford and LIGO Livingston detectors are specifically oriented so that the angles the detectors are at, relative to one another, are precisely compensated for by the curvature of the Earth. This choice ensures that a gravitational wave that appears in one detector will also appear in the other, but the cost is that a gravitational wave that is insensitive to one detector will also be insensitive to the other. In order to get better coverage, more detectors are necessary with a diversity of orientations that are lacking in LIGO Hanford and LIGO Livingston.

But even with up to five detectors, with four independent orientations between them, our gravitational wave capabilities will still be limited in two important ways: in terms of amplitude and frequency. Right now, we have somewhere in the vicinity of ~100 gravitational wave events identified by scientists. But all of them are from relatively low-mass, compact objects (black holes and neutron stars) that have been caught in the final stages of inspiraling and merging together. In addition, they're all relatively nearby, with black hole mergers extended out a few billion light-years and neutron star mergers reaching perhaps a couple of million light-years.

So far, we're only sensitive to the black holes that are around 100 solar masses or under. Again, the reason is simple: gravitational field strengths increase the closer you get to a massive object, but the closest you can get to a black hole is determined by the size of its event horizon, which is primarily determined by a black hole's mass. The more massive the black hole, the larger its event horizon. That means the greater the amount of time it takes for any object to complete an orbit while still remaining outside the event horizon. The lowest-mass black holes (and all of the neutron stars) allow for the shortest orbital periods around them, and even with thousands of reflections, a laser arm that's only 3-4 km long isn't sensitive to longer time periods (shorter frequencies).

That's why if we want to detect the gravitational waves emitted by any other sources, including:

- more massive black holes, like the supermassive ones found at the centres of galaxies,

- less compact objects, like orbiting white dwarfs,

- a stochastic background of gravitational waves, caused by the cumulative sum of all the ripples generated by all of the supermassive black hole binaries whose waves constantly pass us by,

- or the "other" background of gravitational waves: the ones left over from cosmic inflation that still persist throughout the cosmic today, 13.8 billion years after the Big Bang,

We need a new, fundamentally different set of gravitational wave detectors.

The ground-based detectors we have today are limited in the measurement of amplitude and frequency by two factors that cannot be easily improved. The first is the size of the laser arm: if we want to improve our sensitivity or the frequency range that we can cover, we need longer laser arms. With ~4 km arms, we're already seeing just about the highest-mass black holes we can; if we want to probe either higher masses or the same masses at greater distances, we'd need a new detector with longer laser arms. We might be able to build laser arms perhaps ~10 times as long as the current limits, but that's the best we'll ever be able to do because the second limit is set by planet Earth itself: the fact that it's curved along with the fact that tectonic plates exist. Inherently, we can't build laser arms beyond a certain length or a certain sensitivity here on Earth.

Now scientists and researchers are devising another approach that we should begin taking in the 2030s: creating a laser-based interferometer in space. Instead of being limited by either the fundamental seismic noise that cannot be avoided as the Earth's crust moves atop the mantle, or by our ability to construct a perfectly straight tube given the curvature of the Earth, we can create laser arms with baselines hundreds of thousands or even millions of kilometres long. This is the idea behind LISA: the Laser Interferometer Space Antenna, scheduled to be launched in the 2030s.

LISA (Laser Interferometer Space Antenna)

A planned gravitational-wave observatory in space is the Laser Interferometer Space Antenna (LISA). It consists of three identical spacecraft arranged in an equilateral triangle with 2.5-million-km-long sides, more than six times the distance between the Earth and the Moon. LISA is scheduled to launch in the 2030s. LISA will be the first space-based gravitational-wave observatory. The three craft will maintain their formation while following a circumsolar orbit, with test lasers pointing at the other two craft, hence forming a system of Michelson-like interferometers.

The three craft will send laser beams to each other via free-floating golden cubes, each slightly smaller than a Rubik's cube, that are placed inside the craft. The system will be able to measure the separation between the cubes to within the size of a helium atom. Such subtle changes in the distances between the measured laser beams will indicate the presence of a gravitational wave. While ground-based instruments can pick up gravitational waves that have a frequency from a few Hz to a kHz, a space-based mission can detect gravitational waves with frequencies between 10^{-4}–10^{-1} Hz, for example, the coalescence of supermassive black holes.

The orbital motion of the triangle will allow LISA to measure not only the amplitude of the gravitational waves but also the direction of the source. Compared to LIGO, LISA will be mainly sensitive in a lower frequency range, accessing galactic sources of gravitational waves and extreme mass-ratio inspirals.

In June 2017, the European Space Agency (ESA) successfully completed the LISA Pathfinder mission, demonstrating the feasibility of the project. LISA Pathfinder was launched in 2015 on a two-year mission to demonstrate the key technologies required for LISA. LISA Pathfinder consisted of two kg test masses made of gold and platinum that floated freely inside the craft and were separated by 38 cm. The probe also contained a 20 × 20 cm optical bench—containing 22 mirrors and beam splitters—to measure the deviations in their movements to an accuracy of a trillionth of a metre. In April 2016, ESA announced that LISA Pathfinder demonstrated that the LISA mission is feasible.

Figure 8-1.   Leading the way - the LISA Pathfinder spacecraft

LISA Pathfinder made its way to "Lagrange Point 1". Lying about 1.5 million kilometres from Earth in the direction of the Sun, this point in space provides a very stable environment to control the precision of the instruments on the satellite. The launch of LISA Pathfinder marks the 100th anniversary of Albert Einstein's general theory of relativity.

LISA Pathfinder will not be able to detect gravitational waves directly because their impact is so tiny that the test masses would need to be millions of kilometres apart. The experiment will, however, demonstrate that the two independent masses can be monitored as they free-fall through space. The goal is to minimise external and internal disturbances to the point where the position of the test masses would be more stable than the expected change caused by a passing gravitational wave—a change in a distance equal to much less than the size of an atom. The flight of LISA Pathfinder will be a major milestone in the endeavour to observe gravitational waves from space by testing much of the critical technology needed to build a full-scale observatory, such as LISA.

Figure 8-2. The three LISA spacecraft will fly in a triangular formation to detect gravitational waves.

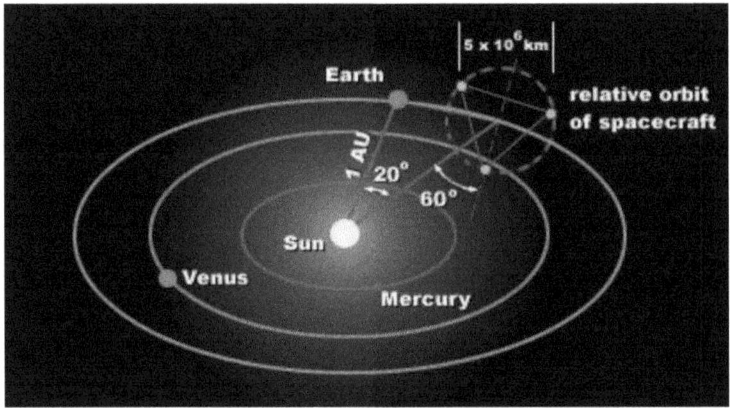

Figure 8-3. Three spacecraft in orbit around the Sun, with a 5 million km baseline. The centre of the triangle formation will be in the ecliptic plane 1 AU from the Sun and 20 degrees behind the Earth. LISA will trail Earth in its orbit around the Sun, which allows it to measure the direction and distance of gravitational wave sources.

LISA will complement traditional astronomical observations based on the electromagnetic spectrum. The detector layout has been carefully chosen to allow observation of most of the interesting gravitational wave sources in the target frequency band between 0.0001 Hz and 0.1 Hz, where the Universe is richly populated by strong gravitational wave sources such as

- supermassive black hole binaries,

- small black holes circling and falling into much more massive black holes (Extreme Mass Ratio Inspirals, EMRIs),

- ultra-compact binaries in our Milky Way,

- stellar-mass black hole binaries years before their merger,

- a stochastic gravitational-wave background, and

- unforeseen and completely unmodelled sources.

With these observations of gravitational waves from the depths of the entire cosmos, LISA will greatly enhance our knowledge about the beginning, evolution, and structure of our Universe. LISA, in many ways, will be the ultimate triumph for what we currently call multi-messenger astronomy: where we can observe light, gravitational waves, and/or particles originating from the same astrophysical event.

The European Space Agency (ESA) has formally approved the start of construction for its space-based gravitational-wave mission. Work on LISA will begin in January 2025 once an industry partner has been chosen to build the craft. LISA, which is estimated to cost €1.5bn, is expected to launch in 2035 and operate for at least four years. On 25 January 2024, ESA's Science Programme Committee formally adopted LISA, deeming that the mission concept and technology are "sufficiently advanced".

EINSTEIN TELESCOPE (ET)

The underground Einstein Telescope will be Europe's most advanced observatory for gravitational waves. The Einstein Telescope (ET) is a proposed underground infrastructure to host a 3rd-generation gravitational-wave observatory. It builds on the success of current second-generation laser-interferometric detectors Advanced VIRGO and Advanced LIGO, whose breakthrough discoveries of merging black holes (BHs) and neutron stars over the past 6 years have ushered scientists into the new era of gravitational-wave astronomy. One of the most promising locations for the Einstein Telescope is the border area of the Netherlands, Belgium, and Germany. Compared to current interferometers, the ET will be able to observe a volume of the Universe approximately one thousand times larger.

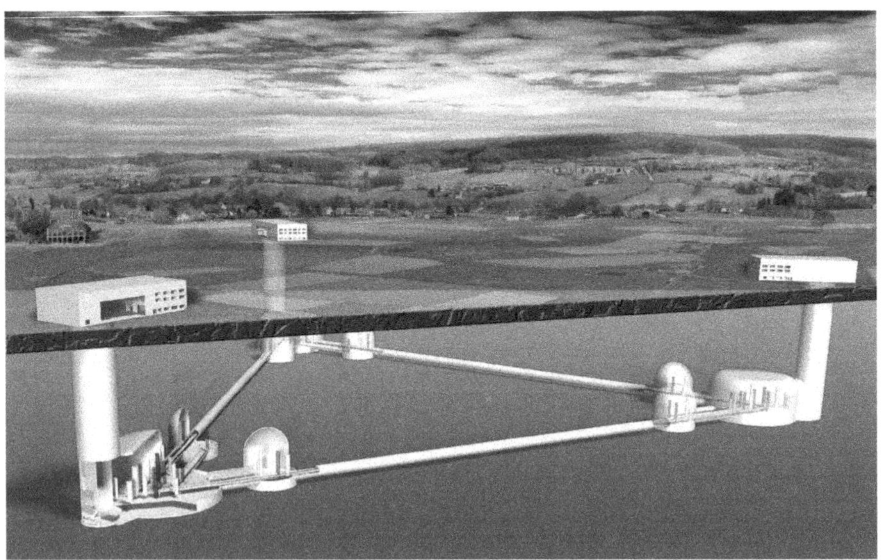

Figure 8-4. This illustration shows the Einstein Telescope. Image Credit: Einstein Telescope, R. Williams (STScI), the Hubble Deep Field Team, and NASA.

The three 10-km tunnels of the Einstein Telescope will be sited 250 to 300 metres underground in order to make undisturbed measurements of gravitational waves. Above ground, hardly anything will be visible in the observatory. Its design calls for 10km long arms, compared to 4 km for LIGO. It will be built underground to reduce seismic noise and background noise from nearby moving objects. The Einstein Telescope will also be cryogenically cooled. It all adds up to increased performance, and it should be able to detect GWs from intermediate-mass black hole mergers.

"Exploiting the ET sensitivity and frequency band, the entire population of stellar and intermediate-mass black holes will be accessible over the entire history of the Universe, enabling us to understand their origin (stellar versus primordial), evolution, and demography," the ET website claims.

The Einstein Telescope will make it possible, for the first time, to explore the Universe through gravitational waves along its cosmic history up to the cosmological dark ages, shedding light on open questions of fundamental physics and cosmology. It will probe the physics near black hole horizons (from tests of General Relativity to quantum gravity), help in understanding the nature of dark matter (such as primordial black holes, axion clouds, and dark matter accreting on compact objects), and the nature of dark energy and possible modifications of General Relativity at cosmological scales.

ET will observe the neutron star inspiral phase and the onset of tidal effects with a high signal-to-noise ratio, providing an unprecedented insight into the interior structure of neutron stars and probing fundamental properties of matter in a completely unexplored regime (QCD at ultra-high densities and possible exotic states of matter). The excellent sensitivity extending to kilohertz frequencies will also allow scientists to probe details of the merger and post-merger phase. ET will operate together with a new innovative generation of electromagnetic observatories covering the band from radio to gamma rays (such as the

Square Kilometre Array, the Vera Rubin Observatory, E-ELT, Athena, and CTA).

With a successful ESFRI (European Strategy Forum on Research Infrastructures) proposal, the project will enter its preparatory phase, which foresees the beginning of construction in 2026 with the goal to start observations in 2035. Two candidate sites are under investigation: one in Sardinia (Italy) and one in the Euroregion Meuse-Rhine. Site-characterisation studies are underway towards a site selection, which is expected in 2024. The evaluation of the sites must consider the feasibility of the construction and predict the impact of the local environment on the detector sensitivity and operation.

COSMIC EXPLORER (CE)

The Cosmic Explorer is another proposed 3rd-generation GW observatory under consideration by the USA. Its conceptual design calls for two separate facilities. One will house two 40 km arms, and the other will have two 20 km arms, each housing a single L-shaped detector.

Cosmic Explorer uses the same L-shaped design as the LIGO detectors, except with 10 times longer arms of 40 km each. This will significantly increase the sensitivity of the observatory, allowing observation of the first black hole mergers in the universe. Cosmic Explorer will push the gravitational-wave frontier to almost the edge of the observable Universe using technologies that have been proven by LIGO during its development. With Cosmic Explorer, scientists can use the Universe as a laboratory to test the laws of physics and study the nature of matter. By observing millions of compact-object mergers across the history of the Universe, Cosmic Explorer will map the populations of neutron stars and black holes across time, bringing new insights into the birth, life, and death of massive stars.

Through exquisite measurements of the interior structure of thousands of neutron stars in mergers, Cosmic Explorer will probe the nature of high-density matter and the strong nuclear force, revealing the nuclear equation of state and its phase transitions in unprecedented detail. The hot, dense remnants of neutron star mergers will map unexplored regions of the quantum chromodynamics phase space. A plethora of multi-messenger observations will help scientists understand the production of the chemical elements that are the building blocks of the Universe and reveal the physics powering short gamma-ray bursts.

While other technological advancements contribute to its sensitivity, its long arms are the key driver, each housing a single L-shaped detector.

Figure 8-5. This illustration shows the Cosmic Explorer. Some of its arms will be 40 km long. Image Credit: Angela Nguyen, Virginia Kitchen, Eddie Anaya, California State University Fullerton; and courtesy of Cosmicexplorer.org

"Sources that are barely detectable by Advanced LIGO, Advanced VIRGO, and KAGRA will be resolved with incredible precision. The resulting explosion in the number of detected sources —up to millions per year—and the fidelity of observations will have wide-ranging impacts in physics and astronomy," the CE website says.

Similar to LISA but sensitive in a frequency band between LIGO and LISA, the Deci-hertz Interferometer Gravitational-Wave Observatory (DECIGO) is a proposed Japanese space-based project.

Pulsar Timing Arrays (PTAs)

At the lowest frequencies, 10^{-9}–10^{-6} Hz, pulsar timing arrays use millisecond pulsars to search for gravitational waves by looking for disturbances in the correlations between arrival times of pulses emitted by objects at different angular separations. Gravitational waves at such low frequencies are expected from galaxy collisions causing mergers of the supermassive black holes at their centres.

Three pulsar timing arrays are currently active: 1. The Parkes Pulsar Timing Array, 2. The European Pulsar Timing Array (EPTA) and 3. The North American Nanohertz Observatory for Gravitational Waves (NANOGRAV). They collaborate within the International Pulsar Timing Array (IPTA) consortium. The Square Kilometre Array (SKA), an intergovernmental international radio telescope project being built in Australia (low frequency) and South Africa (mid-frequency), will contribute significantly to these efforts by detecting many more millisecond pulsars than is currently possible, timing them to very high precision.

The expected number of gravitational wave (GW) events per year with 3rd-generation interferometers, such as the Einstein Telescope (ET) or the Cosmic Explorer (CE), is estimated to be significantly higher than the current generation of detectors like LIGO and VIRGO. Estimates suggest that 3rd-generation interferometers could detect anywhere from tens to thousands of GW events per year, with the Einstein Telescope alone expected to detect around 10,000 to tens of thousands of events per year. These advanced detectors will have improved sensitivity and a larger observation volume, allowing for the detection of a wider range of astrophysical sources. This ushers in a new era of astronomy: one where gravitational waves are no rarer than radio waves and faint signals from collisions in the early Universe echo around the world, loud and clear for all to hear. Overall, gravitational-wave observations are revolutionising our understanding of the Universe and are contributing to scientific knowledge in numerous ways.

The current advancements in technology and understanding of the Universe are leading to a golden age of gravitational wave research, allowing for the study of galaxies, black holes, clusters, and cosmology in greater detail and the exploration of the origins of gravity and the Universe.

Gravitational wave observatories have a bright outlook in the near future. With ongoing advancements in technology and increased collaboration among scientists and institutions worldwide, we can expect several developments:

1. **Improved Sensitivity**: There will be continuous efforts to enhance the sensitivity of existing gravitational wave detectors such as LIGO and Virgo. This will enable them to detect weaker signals from more distant and smaller events, increasing the frequency of detections.

2. **Expansion of the Network**: More gravitational wave observatories are being planned or are in various stages of development globally. For instance, the KAGRA detector in Japan is operational, while the LIGO India project is in progress. The addition of these observatories will improve the ability to localize gravitational wave sources and better understand their properties.

3. **Multi-messenger Astronomy**: Gravitational wave detectors will continue to work in concert with traditional telescopes and other instruments, enabling multi-messenger astronomy. This approach allows scientists to study cosmic phenomena through multiple channels, such as electromagnetic radiation, neutrinos, and gravitational waves, providing a more comprehensive understanding of astrophysical events.

4. **New Discoveries**: As sensitivity increases and more observatories come online, we can anticipate the detection of new types of gravitational wave sources. This could include events involving exotic objects like neutron stars, black holes, or even phenomena predicted by theoretical models but not yet observed.

5. **Technological Innovation**: Research into new technologies, such as quantum squeezing and more advanced laser systems, will further enhance the capabilities of gravitational wave detectors. These technological innovations will push the boundaries of what we can observe and learn about the universe through gravitational waves.

6. Overall, the future of gravitational wave observatories looks promising, with the potential for groundbreaking discoveries and a deeper understanding of the cosmos. These detectors are expected to revolutionise the field of gravitational wave astronomy by significantly increasing the number of detectable events, expanding the range of observable frequencies, and providing valuable insights into various astrophysical phenomena. The multitude of projects aiming to detect gravitational waves over a wide range of frequencies heralds the flourishing of gravitational-wave astronomy within the next few decades. Gravitational waves travel unhindered through the Universe carrying information on its most violent processes,

even those occurring in previously unexplored regions, opaque to photons. In addition, the direct gravitational-wave observations open a path and test the extreme limits of the general theory of relativity, possibly pointing out a way to unite it with quantum mechanics.

CONCLUSIONS

Gravitational wave observations are contributing to scientific knowledge in several ways:

1. Directly observing gravitational waves: Gravitational waves were predicted by Albert Einstein's General Theory of Relativity, and their direct observation confirms the existence of these waves. This observation provides strong evidence for the validity of Einstein's theory and opens up a new window for studying the Universe.

2. Confirming the existence of binary systems: Gravitational-wave observations have confirmed the existence of binary systems, such as binary black holes and binary neutron stars. These observations provide valuable insights into the behaviour and dynamics of these systems, helping us understand how compact astrophysical objects interact.

3. Studying coalescence events: Gravitational-wave observations have allowed scientists to study the coalescence of binary systems, such as black holes and neutron stars, as they merge and form a single object. These events release a tremendous amount of energy in the form of gravitational waves, and studying them helps us understand the physics of extreme astrophysical phenomena.

4. Probing the structure of spacetime: Gravitational waves are ripples in the fabric of space-time itself. By studying these waves,

scientists can gain insights into the structure and properties of spacetime. This research contributes to our understanding of the fundamental nature of the Universe.

5. Advancing astronomy: Gravitational-wave observations provide a new tool for studying the Universe, complementing traditional telescopes and detectors that observe electromagnetic radiation. By combining information from gravitational waves with other forms of radiation, such as radio waves, X-rays, and gamma rays, scientists can gain a more complete picture of astrophysical phenomena.

6. Testing theories of gravity: Gravitational-wave observations can be used to test alternative theories of gravity, beyond Einstein's General Relativity. By comparing the observed properties of gravitational waves with the predictions of different theories, scientists can constrain and refine our understanding of gravity.

7. Opening up new avenues of research: Gravitational-wave observations have opened up new avenues of research in astrophysics and cosmology. They have sparked collaborations between different scientific disciplines and have led to advancements in technology and data analysis techniques.

The Universe is home to some amazing things! Over the past century, scientists have found light from the most distant reaches of the Universe using telescopes on Earth and in space - they have seen the mergers of galaxies and the birth and death of stars. However, orthodox techniques only 'shine light' on a fraction of the Universe. Many of the most interesting things - like black holes - are invisible to optical telescopes. We need a new window: opened not by using light waves but by gravitational waves. Not only will we be able to observe some of the most exotic objects in the Universe with gravitational waves, but we'll also be able to probe the very nature of space and time!

Although we firmly entered the era of gravitational wave astronomy back in 2015, this is a science that is still in its infancy: much like optical astronomy was back in the post-Galileo decades of the 1600s. We only have one type of tool for successfully detecting gravitational waves right now, can only detect them in a very narrow frequency range, and can only detect the closest ones that produce the largest-magnitude signals. As the science and technology underlying gravitational wave astronomy continues to progress, however,

- longer-baseline terrestrial detectors,

- space-based interferometers,

- and increasingly sensitive pulsar timing arrays,

We're going to reveal more and more of the Universe as we've never seen it before. In combination with cosmic ray and neutrino detectors, and being joined by traditional astronomy from across the electromagnetic spectrum, it's only a matter of time before we achieve our first triple bonanza: an astrophysical event where we observe light, gravitational waves, and particles all from the same event. It might be something unexpected, like a nearby supernova, that delivers it, but it might also come from a supermassive black hole merger billions of light-years away. One thing that's certain, however, is that whatever the future of astronomy looks like, it's definitely going to need to include a healthy and robust investment in the new, fertile field of gravitational wave astronomy!

CHAPTER – 9
Gravitational Waves: Timeline

1916: Albert Einstein first proposes the existence of gravitational waves as part of his General Theory of Relativity. Many researchers doubt that they exist at all, believing them to be a mathematical quirk.

1922: English astrophysicist Arthur Eddington's explanation: out of the three types of gravitational waves enunciated by Herman Weyl, only the 'transverse-transverse' wave type propagates at the speed of light in all coordinate systems, so he did not rule out its existence.

1957: A milestone Congress with many scientists and young physicists held in Chapel Hill, North Carolina. Many questions were formulated, and one of them is whether gravitational waves carry energy or not.

1957: Physicists Felix Pirani, Hermann Bondi, and Richard Feynman predicted that gravitational waves carry energy and might be detected by a 'sticky bead argument'. The idea is that if a gravitational wave passed through a stick with a bead on it, it would cause the bead to move back and forth and heat up both the bead and stick due to the friction generated.

1962: Russian scientists Gertsenshtein and Pustovoit published a paper proposing 'interferometers' as a way to detect gravitational waves.

1967: Rainer Weiss (one of LIGO's co-founders) proposes a method that would use laser beams to measure the stretching and squashing of space caused by a passing gravitational wave.

1969: Joseph Weber claims to have detected gravitational waves using a device called a resonant bar detector, but no one can replicate his results.

1974: Rainer Weiss meets physicist Kip Thorne and convinces Thorne that a laser-based instrument would give them the best chance of finding gravitational waves. They started working on the project that would become LIGO.

1978: Russell Hulse and Joseph Taylor provided the first experimental evidence for the existence of gravitational waves by observing two neutron stars orbiting each other (a binary system). They noticed that, rather than remaining in a stable orbit, they were moving closer together (because they were losing energy by emitting gravitational waves) at exactly the rate predicted by Einstein's theory. The discovery earned Hulse and Taylor the 1993 Nobel Prize in Physics.

1984: Kip Thorne, Ronald Drever, and Rainer Weiss founded the LIGO (Laser Interferometer Gravitational-Wave Observatory) Project.

1990s: Construction begins on LIGO—two L-shaped detectors with 4-km-long arms (one in Washington and one in Louisiana), along with gravitational wave detectors in Europe (the VIRGO and GE600 detectors). The idea is that, when a gravitational wave passes through, the arms will lengthen and shorten by a fraction—the precise shift will be measured by lasers travelling along the arms.

2001: LIGO begins operations. At the end of its run in 2010, as expected, LIGO had found no evidence of gravitational waves. LIGO had proved the technology worked—it just needed to be more sensitive.

2007: The VIRGO Laser Interferometer based in Italy, designed to detect gravitational waves, begins operation.

2010: LIGO begins upgrades to become Advanced LIGO. This new facility will be 10 times more sensitive than the old one and includes technology from the UK-German GEO600 detector and from Australia.

2011: VIRGO upgrade commences that will eventually improve the sensitivity by a factor of 10.

2015: In September, Advanced LIGO begins its first engineering and test run. Although only operating at less than half its final sensitivity, it detects its first gravitational wave event on 14th September.

2015: LISA Pathfinder is launched—a testbed mission for the first space-based gravitational wave detector. LISA Pathfinder will test technology for the planned LISA (Laser Interferometer Space Antenna) mission.

2015: On December 26, Advanced LIGO makes a second observation of gravitational waves. This time from two black holes, 14 and 8 times the mass of the Sun, merging into a more massive spinning single black hole 21 times the mass of the Sun.

2016: In February, the LIGO Scientific Collaboration announced that they had indeed detected gravitational waves—GW150914 on September 14, 2015. The waves had been created by two black holes, spiralling in towards each other and merging into a single black hole.

2016: On June 15, the Boxing Day event is announced. This new observation indicates that there is a rich population of binary black holes in the Universe, whose properties are gradually starting to emerge. Gravitational-wave astronomy is no longer a field of single detections but of regular observations. This discovery transforms the LIGO detector into a true astronomical observatory.

2017: The 2017 Nobel Prize in Physics has been awarded to three US scientists, Rainer Weiss (Massachusetts Institute of Technology), Barry Barish (California Institute of Technology), and Kip Thorne (California Institute of Technology), for the detection of gravitational waves. These three scientists have been selected because of their decisive contributions to the development of the LIGO detector and observation of gravitational waves.

2017: The first detection by LIGO and VIRGO of a neutron star-neutron star merger on August 17, 2017, designated GW170817.

2019: GW190521 is the most massive black hole merger yet. This was also the first time that intermediate-mass merging black holes had been detected.

2020: The first and second detection of gravitational waves from a so-called "mixed merger" consisting of a black hole colliding with and swallowing a neutron star. The events designated GW200105 and GW200115 were picked up just 10 days apart on Jan. 5, 2020, and Jan. 15, 2020, respectively.

2021: As of November 2021, the LIGO-Virgo-KAGRA collaboration has detected at least 90 gravitational wave signals from a variety of different events.

The future: Following Advanced LIGO going online in 2017, a 3rd LIGO detector is due for completion in India in 2024. In the 2030s, more sensitive ground-based detectors are foreseen, and LISA will be launched. LISA will extend our capabilities to 'listen' to new kinds of dark phenomena in the Universe.

APPENDIX – I
Pulsar Timing Array (Pta)

A pulsar timing array (PTA) is a set of galactic pulsars that are monitored and analysed to search for correlated signatures in the pulse arrival times on Earth. As such, they are galactic-sized detectors.

Stellar Timekeepers

Pulsars were the first tangible evidence of neutron stars. They don't actually pulsate—they rotate, which causes them to emit a periodic signal. Pulsars are spinning neutron stars that are created when a star explodes as a supernova to leave behind what are the second-most compact objects in our Universe after black holes, in fact, a teaspoon of neutron star matter weighs a staggering 100 million tonnes. Rotating at a rate of up to hundreds of times per second, a pulsar emits a beam of particles and light, including strong radio waves, out of each of its magnetic poles, with the magnetic axis usually precessing around the rotation axis such that each beam sweeps out a conical path in the sky. If a pulsar is oriented such that the solar system lies on this conical path, we can use radio telescopes to detect a "blip" of signal each time its beam comes our way. In fact, this regular signal, which we see at a number of different wavelengths, is our only evidence that pulsars exist.

Some of the fastest pulsars that rotate once every few milliseconds are particularly useful tools because the arrival times of the pulses at our telescopes are so reliably regular that they can be used as extremely precise clocks, even rivalling some atomic clocks. The idea is that if a gravitational wave passes between the pulsar and us, it would alternately stretch and compress the distance that a pulse of light from the pulsar has to travel before it reaches our telescopes. As light moves at a constant speed, the pulses would take a longer or shorter amount of time, respectively, to travel to us, resulting in those blips arriving later or earlier than if there were no gravitational waves at all. The tiny changes in the relative arrival times of the pulses from several pulsars—known collectively as a pulsar timing array—would therefore firmly reveal the presence of a gravitational wave. Or so the thinking goes.

As they rotate, pulsars emit intense electromagnetic radiation that is detected on Earth as regular and precisely timed pulses, which are so regular they are more accurate than an atomic clock. Known pulsars have rates ranging from a few milliseconds to several seconds, depending on the speed of the star's rotation.

Imagine pulsars as cosmic lighthouses meticulously distributed across the galaxy. They emit beams of radio waves with extraordinary regularity. By closely monitoring this network of pulsars, astronomers can accurately measure the arrival of pulses from these distant celestial bodies. This timing technique has the sensitivity to detect minuscule shifts, providing us with vital clues about the influence of gravitational waves.

Pulsar timing arrays monitor millisecond pulsars - rapidly rotating neutron stars that sweep a bright beam of radio waves past Earth several hundred times per second in the Milky Way. As a pulsar rotates, it emits radio waves that sweep by Earth with the period of rotation. The shorter the period, the more and sharper the incident radio wave ticks, so the better the clock. A gravitational wave passing between a pulsar clock and an observer distorts spacetime and causes the signals to arrive either later or earlier, a 'fingerprint' for the detection of the wave. The distortions due to gravitational waves are faint; if they arise from galaxy mergers, they occur with periods of decades.

When the first such millisecond pulsar was found in 1982 by the late University of California, Berkeley astronomer Donald Backer, he quickly realised that these precision flashers could be used to detect the spacetime fluctuations produced by gravitational waves. He coined the term "pulsar timing array" to describe a set of pulsars scattered around us in the galaxy that could be used as a detector.

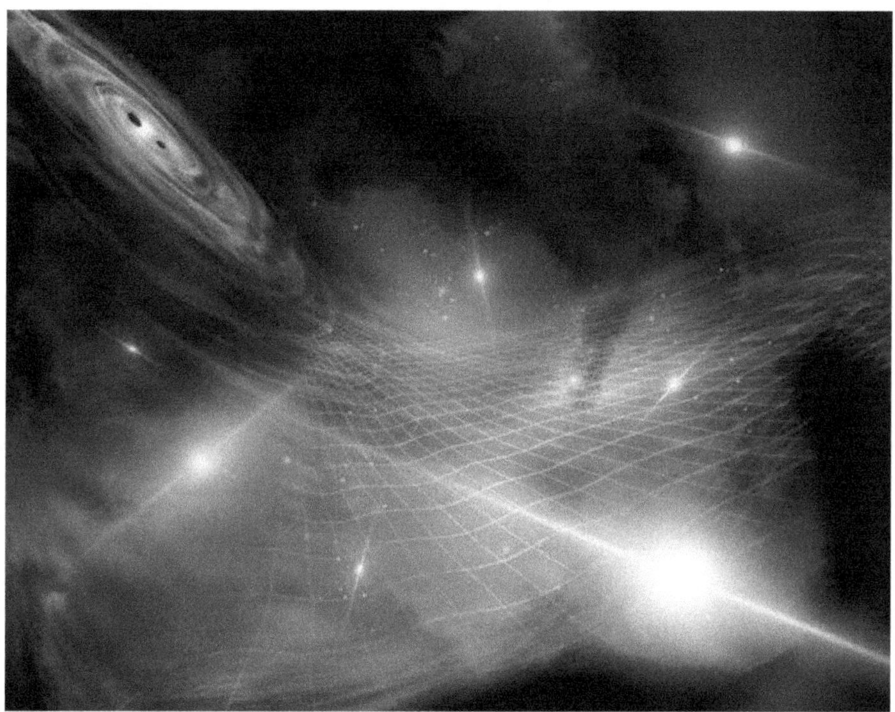

Figure 1.   Artist's interpretation of an array of pulsars being affected by gravitational ripples produced by a supermassive black hole binary in a distant galaxy. Credit: Aurore Simonnet for the NANOGrav Collaboration.

Unlike the fleeting high-frequency gravitational waves seen by ground-based instruments like LIGO (the Laser Interferometer Gravitational-Wave Observatory), this continuous low-frequency signal could be perceived only with a detector much larger than the Earth. To meet this need, astronomers turned our sector of the Milky Way Galaxy into a huge gravitational-wave antenna by making use of exotic stars called pulsars.

Extragalactic gravitational waves wash over all the pulsars in the Milky Way. Because the pulsars are independent, and each has its own timing and its own interstellar medium, the main giveaway for detecting a gravitational wave is a correlated signal between the pulsars. Pulsars do their own thing, and it's hard to dig out a tiny signal from a large number of sources of noise. The main way to overcome this is by timing an array of pulsars. Each experiment keeps tabs on around 50 pulsars. Often, an individual astronomer (in the PTA) is responsible for a specific clock and monitors more than two at a time.

The International Pulsar Timing Array (IPTA) is a global consortium that includes gravitational wave detectors like the North American Nanohertz Observatory for Gravitational Waves (NANOGrav), the European Pulsar Timing Array (EPTA), the Indian Pulsar Timing Array Project, and Australia's Parkes Pulsar Timing Array (PPTA). All of them cooperate through the International Pulsar Timing Array (IPTA). The NANOGrav collaboration monitored 68 pulsars in total, some for 15 years, and employed 67 scientists in the current analysis. The NANOGrav data allow several other inferences about the population of supermassive black hole binary mergers over the history of the Universe. This discovery, based on 15 years of data, aligns with predictions of General Relativity and offers insights into supermassive black holes and galaxy mergers.

Pulsar Timing Arrays (PTAs) serve as our unique lens into the Universe. They allow us to "listen" to the cosmic symphony and unravel some of the most profound mysteries of the cosmos. Pulsar timing involves meticulously measuring the times of arrival (TOAs) of pulses emitted by pulsars. Any deviation in these TOAs can indicate changes in the medium through which the signals pass, including spacetime distortions caused by gravitational waves.

Radio telescopes capture these signals, and the data undergo extensive analysis to create a comprehensive "timing model" for each pulsar. This model considers various factors, such as the pulsar's spin period, spin-down rate, and location. Computer algorithms then compare the observed TOAs with the predicted TOAs from the timing model.

If gravitational waves have traversed the space between the pulsar and Earth, they will cause a correlated change in the TOAs of a group of pulsars—this correlation is what PTAs diligently seek. The pulsar timing array method is most sensitive to low-frequency gravitational waves and so requires years of observations.

Future investigation of this gravitational wave "cosmic hum" will feed into scientists' understanding of how the Universe evolved on the largest scales, providing information about how often galaxies collide and what drives black holes to merge. In addition, gravitational ripples of the Big Bang itself may make up some fraction of the signal, offering insight into how the Universe itself was formed. These results even have implications at the smallest scales, placing limits on what kind of exotic particles may exist in our Universe.

APPENDIX – II

Neutron Star Merger

Einstein's Theory of Relativity posits that if you have no rest mass as you travel through the vacuum of space, you are compelled to travel exactly at the speed of light. This is true for all massless particles, like photons and gluons. It is approximately true for particles like neutrinos, whose mass is tiny compared to their kinetic energy, and should also be true for gravitational waves. Even if gravity isn't inherently quantum in nature, the speed of gravity should be equal to the speed of light if our current laws of physics are correct.

On August 17, 2017, the signal from an event GW170817, which occurred 130 million light-years away, finally arrived here on Earth. From somewhere within the distant galaxy NGC 4993, two neutron stars had been locked in a gravitational dance where they orbited one another at speeds that reached a significant fraction of the speed of light. As they orbited, they distorted the fabric of space through which they travelled.

Whenever masses accelerate through curved space, they emit tiny amounts of invisible radiation that's invisible to all telescopes: gravitational waves. These gravitational waves behave as ripples in the fabric of spacetime, carrying energy away from the system and causing their mutual orbit to decay. At a critical moment in time, these two stellar remnants spiralled so close to one another that they touched, and what followed was one of the most spectacular scientific discoveries of all time.

As soon as these two stars collided, the gravitational wave signal came to an abrupt end. Everything that the LIGO and VIRGO detectors saw was from the inspiral phase up until that moment, followed by total gravitational wave silence. According to our best theoretical models, this was two neutron stars inspiraling and merging together, likely resulting in a remarkable end result: the formation of a black hole.

But then it happened. After the gravitational wave signal ceased, 1.7 seconds later, the first electromagnetic (light) signal - gamma rays arrived, which came in one enormous burst. From the combination of gravitational wave and electromagnetic data, scientists were able to pinpoint the location of this event better than any gravitational wave event ever: to the specific host galaxy in which it occurred, NGC 4993.

Over the coming weeks, light began to arrive in other wavelengths as well, as close to 100 professional observatories throughout the world monitored the spectacular afterglow of this neutron star merger.

Figure 1. *For the 2017 neutron star-neutron star merger, an electromagnetic counterpart was robustly seen immediately, and follow-up observations, such as this Hubble image, were able to see the afterglow and remnant of the event.*

On the one hand, this is remarkable. We had an event occur some 130 million light-years away: far enough away that light took 130 million years to travel from the galaxy where it occurred to our eyes. For all that time, from then until the present, both the light and the gravitational waves from this event were journeying through the Universe, travelling at the only speed they could—the speed of light and the speed of gravity, respectively—until they arrived at Earth after a journey of 130 million years. First, the gravitational waves from the inspiral phase arrived, moving the mirrors on our gravitational wave detectors by an incredibly small amount: less than a ten-thousandth of the size of an individual proton. And then, just 1.7 seconds after the gravitational wave signal ended, the first light from the event arrived as well.

Figure 2.   *An illustration of a very high-energy process in the Universe: a gamma-ray burst. These bursts can arise when two neutron stars merge, and one was detected just 1.7 seconds after the gravitational wave signal from GW170817 ceased. (NASA / D. BERRY NASA / D. BERRY).*

The key is to think about the objects that are merging, the physics at play, and the signals they're likely to produce.

During an inspiral and merger of two neutron stars, a tremendous amount of energy should be released, along with heavy elements, gravitational waves, and an electromagnetic signal. Up until these two neutron stars touched, there was no "extra" light produced. They simply shone as neutron stars do: faintly, at high temperatures but with tiny surface areas, and completely undetectable with our current technology from 130 million light-years away. Neutron stars aren't like black holes; they aren't point-like. Instead, they're compact objects—typically somewhere between 20 and 40 km across—but denser than an atomic nucleus. They're called neutron stars because they're about 90% neutrons by composition, with other atomic nuclei and a few electrons at the outer edge.

When two neutron stars collide, there are three possibilities that can result. They are:

1. Formation of another neutron star, if the total mass is less than 2.5 times the mass of the Sun,

2. Formation of a new neutron star briefly, which then collapses into a black hole in under a second if the total mass is between 2.5 and 2.8 solar masses (dependent on the neutron star's spin),

3. Formation of a black hole directly, with no intermediate neutron star, if the total mass is greater than 2.8 solar masses.

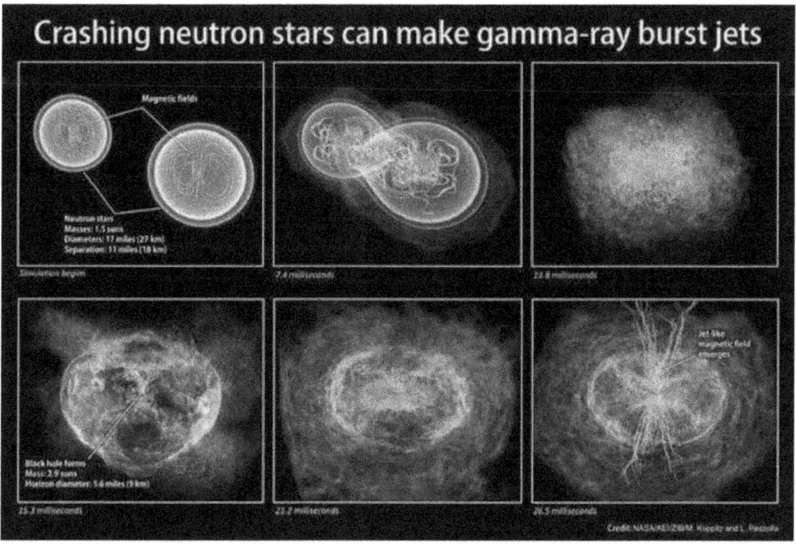

Figure 3.   *We knew that when two neutron stars merge, as simulated here, they can create gamma-ray burst jets, as well as other electromagnetic phenomena. But perhaps, above a certain mass threshold, a black hole is formed where the two stars collide in the second panel, and then all the additional matter and energy gets captured, with no escaping signal. (NASA / Albert Einstein Institute / Zuse Institute Berlin / M Koppitz and L Rezzolla)*

From the gravitational wave signal that arose from this event, officially known as GW170817, the results show that this event falls into the second category: the merger and post-merger signal existed for a few hundred milliseconds before disappearing entirely in an instant, which indicates that a neutron star formed for a brief time before an event horizon formed and engulfed the entire thing.

But nevertheless, light still got out and there were three possibilities that we could think of.

1. Immediately, as soon as the neutron stars touch, processes occur on their surfaces.

2. Only after the material is ejected, it collides with any surrounding material and produces light from that.

3.   Or from the interior of neutron stars, where reactions generate energy that is only emitted once it propagates to the exterior.

In each scenario, gravitational waves travel unperturbed once the signal is generated, but light takes an extra amount of time to get out. If it's the first option, and neutron star mergers generate light as soon as they touch, the light gets emitted immediately and therefore must be delayed by passing through the environment surrounding the neutron star. That environment must be rich in matter, as each fast-moving neutron star, with charged particles on its surfaces and intense magnetic fields, is bound to strip and eject material from the other one.

If it's the second or 3rd option, merging neutron stars generate light from their mergers, but that light only gets emitted after a certain amount of time has passed: either for ejected material to smash into the circumstellar material or for the light generated in the neutron star interiors to reach the surface. It's also possible, in either of these cases, that both "delayed emission" and "slowed arrival by surrounding material" are at play.

Any of these scenarios could easily explain the 1.7s delay of light's arrival with respect to gravitational waves. But on April 25, 2019, scientists observed another neutron star-neutron star merger in gravitational waves, which was more massive than GW170817. No light was emitted of any type, discounting the first scenario. It looks like neutron stars don't generate light as soon as they touch. Instead, the emission of light comes after the emission of gravitational waves.

With the studies on NS-NS binary mergers, now it's definitively shown that a large fraction of the elements in our Universe—including gold, platinum, iodine, and uranium—arise from these neutron star mergers. But not, perhaps, from all neutron star mergers; perhaps it's only the ones that don't immediately form a black hole. Either ejected material or reactions in the neutron star's interior are required to produce these elements, and hence, the light associated with a Kilonova explosion. That light is only produced after the gravitational wave signal has ended and may further be delayed by having to pass through the circumstellar material.

Bibliography

1. "Astrophysical Implications of the Binary Black Hole Merger GW150914" by Abbott, B.P. et al., The Astrophysical Journal Letters, 818: L22 (15 pp), 2016.

2. "The First Sounds of Merging Black Holes" by Emanuele Berti, Phys. aps.org, 11 February 2016, Physics 9, 17.

3. "Observation of Gravitational Waves from a Binary Black Hole Merger" By B.P. Abbot Etal Physics Review Letters, PRL 116, 061102-pp: 1-15, Feb 2016.

4. "Gravitational Waves Discovered from Colliding Black Holes" By CLARA MOSKOWITZ SPACE: Scientific American, Feb 11, 2016.

5. "Direct Observation of gravitational waves" LIGO Educator's Guide http://www.ligo.org/updates https://dcc.ligo.org/LIGO-P1600015/public

6. "Gravitational Waves: Everything You Need to Know" By Ben Gilliland, Science and Technology Facilities Council (STFC). Published in Big Science at STFC on 20th January 2023.

7. "A brief history of Gravitational Wave research" By Chiang-Mei Chen, et al. PACS number: 04; 20.Cv, 04.20Fy arXiv:1610.08803v2 (gr-qc) 22 November 2016.

8. "Brief History of Gravitational Waves" By Jorge L. CERVANTES COSTA arXiv: 1609.09400 (physics.hist-ph) 26th September 2016

9. "LIGO and Detection of Gravitational Waves" By Barry C. Barish and Rainer Weiss Physics Today 52, 10, 44 (1999) http://dx.doi.org/10.1063/1.882861

10. "Hulse and Taylor win Nobel Prize for Discovering Binary Pulsar" By Bertram Schwarzschild Physics Today 46(12), 17-19 (1993) https://doi.org/10-1063/1.2809120

11. "The Laser Interferometer Gravitational Wave Observatory and the first direct observation of gravitational waves" Scientific Background on the Nobel Prize in Physics 2017, The Royal Swedish Academy of Science, Nobel Prize.org

12. "Making Waves: Scientists 'See' Gravitational Waves" By T.V. Venkateswaran, Cover Story, Science Reporter, March 2016.

13. "Popular Science Background" The Nobel Prize in Physics 2017, Nobel Prize.org

14. "Gravitational Wave Astronomy: A New Window to the Universe" By P. Ajith and K. G. Arun, Resonance; October 2011, pp. 922-932.

15. "Here Come the Waves" By DAVIDE CASTELVECCHI Scientific American: Space Physics, Issue No. 2, June-July 2018

16. "When Black Holes Merge" By LAMBODARA MISHRA SCIENCE REPORTER, AUGUST 2012, pp. 46-48.

17. "Gravitational Wave Observatory" By Y. BALA MURALI KRISHNA SCIENCE REPORTER, November 2012, pp. 44-48.

18. "Gravitational Waves" By Kip S Thorne Proceedings of the 94 Summer Study on Particle and Nuclear Astrophysics and Cosmology, 1995.

19. "Gravitational Wave search resumes after 3 Years and Lots of Headaches" By Daniel Garisto Scientific American, 23 May 2023 www.scientificamerican.com

20. "Gravitational Waves: A New View of the Universe" By Priyanka Roy Chowdhary and Sukhdev Sahoo I.A.P.T. Bulletin, August 2018, pp. 208-213.

21. "Pulsar Timing Arrays are poised to reveal gravitational waves" By TONI FEDER Physics Today, 70(7), 2019, pp. 26-28

22. "An Ear to the Big Bang" By Ross D. Anderson Scientific American, October 2013, pp. 26-33

23. "LIGO comes to India" By Chetna Krishna Symmetry Magazine, 16 February 2024 http://www.symmetrymagazine.org/article/ligo-comes.to-india

24. "ASTRONOMY'S OTHER SENSE: GRAVITATIONAL WAVES" BY DAN LAMBETH SPACE AUSTRALIA, 23 SEPTEMBER 2020.

25. "The Current Status and Future Prospects of KAGRA, The Large-Scale Cryogenic Gravitational Wave Telescope Built in the Kamioka Underground" By Homare Abi et al. Galaxies 2022, 10(3), 63 https://doi.org/10.3390/galaxies10030063

BOOK SOURCES

1.   "Multi-Messenger Astronomy"

    By Imre Bartos and Marek Kowalski

    IOP SCIENCE, IP Address: 157.48.78.179

    ISBN 978-0-7503-1369-8 (e-book).

2.   "LIGO-INDIA"

    Indigo: Indian Initiative in Gravitational Wave Observations

    http://www.gw-indigo.org

    10 November, 2011

3.   "Gravitational Waves: A New Window to the Universe"

    Ajith Kumar and Pushpa Khare

    Springer Nature Singapore Pvt Ltd, 2020, XXVI, pp. 161.

4.   "Gravitational Waves"

AMBER L STUVER

IOP Science, iop.org

IP Address: 103.24.124.46, March 23, 2022.

5.   "RIPPLES IN Spacetime: Gravitational Waves and the Future of Astronomy" – Kindle Edition

SCHILLING GOVERT

https://lecn.loc.gov/2017007571

6.   "DISCOVERING GRAVITATIONAL WAVES" – Kindle Single

JOHN GRIBBIN